THE HUMAN FACE O1

General Nick Carter
With compliments
Jim Storr

OTHER BOOKS IN THE SERIES

Militarised Landscapes
British Army in Battle and Its Image 1914–18
Red Coat, Green Machine

Jim Storr has a portfolio career in the defence sector. His expertise lies in the linkage of military concepts, doctrine, history, technology and human behaviour. His main areas of work are consultancy, teaching, writing and research. He is a frequent speaker at international conferences.

Dr Storr was an army officer for 25 years. He served in Britain, Germany, the Falklands, Northern Ireland, Cyprus and Canada. He wrote the British Army's current high-level tactical doctrine, and its analyses of warfighting operations in Iraq in 2003 and in Northern Ireland since 1969.

He holds a PhD, a master's degree in defence technology, and a first degree in civil engineering. He is a visiting fellow of Cranfield and Birmingham Universities.

Series Editor: Gary Sheffield, Professor of War Studies, University of Birmingham.

Series Associate Editor: Dan Todman, Senior Lecturer, Queen Mary, University of London.

The Human Face of War

Jim Storr

BIRMINGHAM WAR STUDIES SERIES
Series edited by Professor Gary Sheffield
and Dr Don Todman

continuum

Continuum UK, The Tower Building, 11 York Road, London SE1 7NX
Continuum US, 80 Maiden Lane, Suite 704, New York, NY 10038

www.continuumbooks.com

First published 2009

British Library Cataloguing-in-Publication Data
A catalogue record for this book is available from the British Library.

ISBN 978 1 4411 8750 5

Typeset by Pindar NZ, Auckland, New Zealand
Printed and bound by MPG Books Ltd, Cornwall, Great Britain

Contents

Figures

To
Private Thomas Atkins
No 6 Troop, 6th Dragoons
and others like him

Abbreviations

AAA	Anti-Aircraft Artillery.
AFV	Armoured Fighting Vehicle.
ATGW	Antitank Guided Missile.
BAOR	British Army of the Rhine.
BATUS	British Army Training Unit, Suffield (Canada).
BEF	British Expeditionary Force.
CAS	Close Air Support.
CGS	Chief of the General Staff.
CIGS	Chief of the Imperial General Staff.
C.-in-C.	Commander in Chief.
CO	Commanding Officer.
COS	Chief of Staff.
CP	Command Post.
CPD	Continuing Professional Development.
CSS	Combat Service Support: medical and logistic support.
C2	Command and Control.
DERA CDA	Defence Evaluation and Research Agency – Centre for Defence Analysis.
DSO	Distinguished Service Order.
EBO	Effects-Based Operations.
EW	Electronic Warfare.
FIBUA	Fighting in Built-Up Areas.
FM	Field Marshal.
FOO	Forward (artillery) Observation Officer.
GI	General Infantryman.
GOC	General Officer Commanding.
GST	General Systems Theory.
HA	Historic Analysis.
HB(A)	Army Historical Branch. The name of a former branch of the British MoD.
HE	High Explosive.
HESH	High Explosive, Squash Head.
HQ	Headquarters.

IR	International Relations.
LMG	Light Machine Gun.
MBTI	Myers–Briggs Type Indicator.
MC	Military Cross.
MG	Machine Gun.
MoD	Ministry of Defence.
NCO	Non-Commissioned Officer.
NEC	Network-Enabled Capability.
OA	Operational Analysis.
OAS	Offensive Air Support.
OODA	Observation, Orientation, Decision and Action – as in 'the OODA Loop'.
PWO	Principal Warfare Officer.
RHA	Royal Horse Artillery.
SAM	Surface-to-Air Missile.
TA	Territorial Army.
TOW	US BGM 71 long-range ATGW system. The acronym stands for 'Tube launched, Optically tracked, Wire guided', but the system is universally referred to as 'TOW'.
VC	Victoria Cross.
VE	Victory in Europe. VE Day was 8 May 1945.
2ic	Second in Command.

Foreword

This is 'a book about warfare; the conduct of war'. It is neither a comfortable subject nor a comfortable read. Its author continually challenges some of the assumptions which surrounded him for almost 25 years of soldiering. He is remorseless in his slaughter of loose thought, unsubstantiated argument and ill-defined belief, and suggests that part of the doctrine with which he was closely involved in the latter stages of his career is characterized by lack of intellectual rigour and sloppy use of English. For instance, he says of the 'Boyd Cycle' – the process of Observation–Orientation–Decision–Action sometimes called the OODA loop – that 'to generalize about formation-level C2 from aircraft design is tenuous, to say the least'. He warns us that we should expect more from military thought than a jumble of poorly described phenomena shot though by apparent paradox, and stresses that, whatever the virtues of classical theorists like Carl von Clausewitz or Sun Tzu, we need to understand them better if we can hope to assess their contemporary relevance. 'In which other discipline so vital to man's existence', he wonders, 'would we grant almost divine reverence to one long-dead German?' His clear thought was much admired by his colleagues, though his determination to apply the humane killer to the occasional sacred cow led to one senior officer to describe him as 'about as popular as an Old Testament prophet'.

One of the author's many strengths is his ability to place warfare in a wider intellectual context, with frequent references to engineering, medicine and General Systems Theory. His experience urges him to strike practical warnings: 'so it may be true', he tells us, 'and it might appear simple. That does not make it easy.' Many of the truths he identifies are indeed fundamental. He reminds us that 'we should attack the enemy's will with speed and surprise', but points out that, paradoxically, the process of making decisions and issuing orders from large and complex modern headquarters has actually slowed their pace of activity. Plans which are 'inherently successive rather than simultaneous', tend, even when they succeed, to be slower and more costly for the victor than schemes which generate genuinely simultaneous activity. He emphasizes the importance of the prompt focusing of combat power, and points to the close relationship between military structures and their ability to generate this power on the battlefield. Tactical decisions must be made quickly, and need to be 'about right': good-ish decisions made speedily tend to beat perfect decisions made late.

Jim Storr agrees with Sir Michael Howard that doctrine is often a poor guide to the opening clashes of major conflicts: what counts is the ability to transform it in a timely and effective manner. The British Army has a patchy track-record in its selection of senior commanders at the beginning of big wars, and thus finds itself needing to change both doctrine and its leadership at a time of urgent crisis. I yield to no one in my admiration for the army, but share the author's suspicion that its culture may not yet have evolved to the point where his concerns about the performance of its leaders in conflict are simply a matter of benevolent historical reflection.

Much of the book is devoted to war-fighting, and the author recognizes that there is a powerful line of argument, recently encapsulated in General Sir Rupert Smith's *The Utility of Force*, that war between states is now a thing of the past. In contrast, he sympathizes with Professor Colin Gray's view that 'timeless reasons of fear, honour and interest' will continue to make 'old-fashioned' war a feature of the future. Indeed, it is worth emphasizing that some aspects of current operations in both Iraq and Afghanistan are decidedly old-fashioned rather than postmodern. If the Americans and their coalition partners lose it is far more likely to be because they have forgotten ancient lessons (not least the fundamental need for a cohesive strategy, and the inability of tactics and technology to compensate for its deficiencies) rather than failed to learn new ones.

In the last analysis this is a book about the way that people fight. Storr suggests that we should look at war in general, and combat in particular, as a human rather than a technical phenomenon. The essence of his argument is stark:

> We should base our understanding of combat on a fundamental premise: that fighting battles is basically an assault on the enemy army as a human institution. Combat is an interaction between human organizations. It is adversarial, dynamic, complex and lethal. It is grounded in individual and collective human behaviour, and fought between organizations that are themselves complex. It is not determined, hence uncertain, and evolutionary. Critically, and to extent which we currently overlook, combat is a *fundamentally* human activity. This, then, is our new paradigm.

I urge those readers who find themselves in disagreement with parts of his argument to recognize that the author shares the conviction, which unites so many of us, that the men (and these days, women too) who do the business of battle deserve the very best that we can give them. Arguments are not necessarily wrong because we dislike them, or conclusions false because we inherently shy away from them. Think beyond the grubby little orthodoxies of our own time. Too often we worry most about weapons and equipment. Yet I can think of no time in the whole of its long and not undistinguished history that the British Army had quite enough of the very latest weaponry, and to complain that it does not do so today is to miss at least part of the point. Many of the decisive battles of history had an outcome that would not have been predicted by simple spear-counting. Intellectual and moral qualities, often hard to evaluate in an age obsessed with quantification, are of fundamental importance, and yet we are too often inclined to allow technology to trump thought. This book tells us why thought matters, and why we should recognize that the human being lies at the centre of combat. Whether that combat takes place as part of a conventional war, or in the context of that asymmetric conflict which seems to characterize our times, is not the issue. It will, as Jim Storr so firmly assures us, remain a fundamentally human activity, as we forget that at our peril.

Richard Holmes

Series Editors' Preface

War Studies is an influential, popular and intellectually exciting discipline, characterized by the broad range of approaches it employs to understand a fundamental human activity. It has the history of war at its heart, but goes beyond operational military studies to draw on political, cultural and social history as well as strategy, literature, law, political theory, economics and social science. Birmingham War Studies celebrates this diversity by publishing examples of the discipline at its best, bringing the best academic scholarship on war to a wide audience. The remit of the series is deliberately and ambitiously wide, ranging across periods, methodologies and geographic regions. In this, it takes its cue from the treatment of the subject by the War Studies Group (WSG) within the Department of Modern History at the University of Birmingham. The launch of the series as a joint venture between Continuum and the WSG marks the latter's emergence as a major force in the field, both in the UK and internationally.

That of the first three books to be published, one deals with military theory and doctrine, and another with social anthropology, indicates that, while 'traditional' military history has its place in Birmingham War Studies, the series' scope is very broad. The high quality of research, analysis and expression will make these books required reading for all those interested in the study of war in general as well as in the topic covered by each volume. In each case, the methodologies used in these specific studies will drive the field as a whole forward.

Jim Storr's *The Human Face of War* is therefore an absolutely appropriate point from which to take off. Founded upon a deep knowledge of military history but drawing upon many disciplines, this book presents a challenging, iconoclastic, caustic and persuasive critique of contemporary military thought. Dr Storr writes as an insider. He is a former Regular infantry officer in the British Army and he blends evidence drawn from history and earlier military thought with his own experience as a soldier, which included a significant period writing military doctrine. The result is a masterly reassessment of the subject that deserves the widest dissemination in both academic and military circles, especially as it has much to say about a subject, conventional warfare, that is currently unfashionable, with conflict at the other end of the spectrum of organized violence, such as terrorism and counter-insurgency, currently monopolizing thinking and writing. The level of analysis and breadth of reference which Dr Storr employs make this an extremely thought-provoking book, which has much to say about the nature of war to all those who study it. This is a history of thought (and the lack of thought), but it is also an explanation of how armies function and soldiers fight. It is bound to be controversial, but controversy seems to go with the territory. This book will undoubtedly stimulate debate and engender further thought, and will thereby achieve one of its primary aims. *The Human Face of War* is an impressive and original piece of work. No future writer on military thought and doctrine, in either an historical or contemporary context, can afford to ignore it.

Gary Sheffield, editor, Birmingham War Studies
Dan Todman, associate editor, Birmingham War Studies
Birmingham and London, March 2009

Introduction

This book is about warfare: the conduct of war. Briefly, it considers what war is like, and then discusses the consequences for armies, both on and off the battlefield. Its main tenet is that, although war currently appears to be dominated by technology, warfare is fundamentally a human issue. It focuses on major, state-on-state war – what some people call 'regular' war, and others 'old-fashioned' war.

That may seem strange. What the man on the street sees of war today is highly coloured by the events of the so-called 'Global War on Terrorism'. Iraq, Afghanistan and a seemingly world-wide, networked fundamentalist campaign of terror dominate national and international events. Is old-fashioned war a thing of the past? Esteemed writers – amongst them a respected retired general and a well-known Israeli academic – have suggested that old-fashioned, state-on-state war is a thing of the past. Future wars will be fought 'amongst the peoples'. The existence of nuclear weapons renders interstate conflict less likely. Indeed 'war no longer exists', not least because the 'will of the people' cannot ultimately be coerced.[1]

Really? More than 60 years after the first use of nuclear weapons, a tiny number of countries possess them – at most a dozen, perhaps 5 per cent of the world's nations. Only about 30 nations possess the nuclear power stations which produce the fissile material needed to make them. Just two more states – Iraq and North Korea – might have usable nuclear weapons within a decade. One state – Libya – has recently given up trying to acquire them. At that rate, it will be many years before a significant proportion of the world's 200 or so states have nuclear weapons. Nuclear deterrence will probably not be a major factor in the relationships between most countries for a long time. Furthermore, any attempt to acquire nuclear weapons might just prompt a war to prevent that happening.

Regrettably, the will of the people can be coerced, for decades at a time. If that were not so, many of the more unpleasant regimes of the twentieth century would not have endured as long as they did. So one might reasonably doubt the underpinnings of this 'postmodern' thesis. Western nations, and particularly Western European nations, may like to assume that peace and freedom are inevitable. That is to overlook the fact that such luxuries have often been bought at the price of much blood. Peace and freedom are ultimately sustained by the ability to pay that price again if necessary. The right to live in freedom implies a duty to protect oneself against attack.

Old-fashioned war may not be particularly prevalent at present. That may lull us into thinking that it isn't likely to be. As Professor Colin Gray put it: 'the unsound belief that major war is obsolete, or obsolescent, rests on nothing more solid than superficial trend spotting. It is scarcely a triumph of perceptive scholarship . . .'[2] One can think of four possible reasons why major war is not currently prevalent. First, after two of the most destructive wars in history, Western Europe has convinced itself of the need to live at peace within its own borders. Secondly, after 1991 no country will risk war against the USA (a lesson strongly reinforced in 2003). Thirdly, after 1973 no Arab country will risk war against Israel. Lastly, Western states are quite proficient at waging war against non-Western states. Several of those factors serve as a tangible deterrent. It would, however, be naïve to assume that any or all of them will apply indefinitely. Thus studying how best to wage regular, old-fashioned war is still critically important.

Professor Gray suggests that 'old-fashioned' state-on-state war will be a feature of the future for timeless reasons of fear, honour and interest.[3] I find his arguments to be more persuasive than the alternative viewpoint. Gray's perspective is grounded in a long, hard look at history. History cannot predict the future, but it does seem to be our best guide to it. It has limitations, however, some of which are discussed in this book.

This is a book about warfare: the conduct of war. There aren't many others. The great classics are Clausewitz and Sun Tzu. But in which other discipline so vital to man's existence would we grant almost divine reverence to one long-dead German? Fuller and Liddell Hart made significant contributions, but in somewhat limited areas. Other than that, there are specific works about 'what future wars will be like', such as General Rupert Smith's *The Utility of Force* and Gray's *Another Bloody Century*. There are other relatively narrowly focused works such as Bill Lind's *Maneuver Warfare Handbook*[4] or Edward Luttwak's work on strategy.[5] There are a few others, some of which are quite good. Professor Martin van Creveld's books on command and logistics in war are certainly among them.[6] But in general, there are few works in this area.

Conversely, there are any number of works of military history. They tell us what *has happened*. But they don't tell us *how to do it*. Equally there are any number of works on international relations, where 'strategy' is effectively an extension of politics and diplomacy. They may tell us why we wage war, or the consequences of doing so. But again, they don't generally tell us how to do it.

The Human Face of War is about military theory. That is an important issue. At times the fate of countries and even whole continents rests on warfare, and billions of pounds are spent each year preparing for, or waging, war. Yet military theory is a curiously underdeveloped discipline. We should ask some very basic questions, and then develop a better way of looking at war, at how armies should fight and how they should be organized.

Perhaps surprisingly, much current military theory is incoherent, inconsistent and poorly written. The underlying philosophy seems flawed. If, as Clausewitz considered, fighting is the basic currency of war, we should take a long, hard look at what combat – fighting – is like. It appears to be vastly complex and to verge on chaos. To understand complexity one needs a grasp of the basics of complexity theory and the fundamentals of General Systems Theory. If we arm ourselves with such tools, we can better understand battlefield phenomena such as destruction, protection, participation in combat and the social cohesion of an armed force. We can begin to see how they relate to each other. Other human issues, such as shock and surprise, are central to fighting but are generally neither well explored nor described.

Once we have considered those issues we are better placed to understand what actually happens on a battlefield, and what a commander can do to win. For commanders are paid to win: in war, nobody gets paid to come second. That leads into a consideration of tactics – the processes of combat – and organization. A recurrent theme in this discussion is that tactical success comes when, and only when, the enemy believes himself to be beaten. How should one make that happen? What kinds of tactics should be used, and how should forces be organized to conduct them? How should those forces be commanded on the battlefield?

The individual plays a key role in warfare. Some commanders have a dramatic impact on the outcome of battle, and some armies seem to generate such commanders more consistently than others. Similarly, some individual fighters can have a hugely disproportionate effect: snipers, tank commanders, fighter-pilots and submarine commanders are all good examples. Can armed forces produce such experts on a reliable basis? Should they try?

Armed forces are composed not just of individuals, but groups of them. That raises issues of sociology and culture. Armies need to exist from one war to the next without doing the job they are designed to do: fight. They don't always get that right, largely for organizational reasons, and the consequences may visit them at the beginning of the next war. The type of people who achieve high rank also affects that process. Examining the personalities of senior commanders reveals two main types – authoritarians and autocrats – who differ in important ways. Some seem almost predestined to fail in war. So what is needed is not just a body of theory as to how to fight; but also how to organize armies in peacetime to fight and win when needed. Organization, doctrine, training policy and issues which affect social cohesion and career progression are all relevant factors.

Overall, war is not as it has generally been described. Combat – and to that extent war – is complex, adversarial, lethal, dynamic and evolutionary. It is fundamentally human, and waged between complex human organizations. It is uncertain and risky. Critically, it is not determined by technology alone. These statements are, in effect, the core of a new way of understanding war: a new paradigm for warfare. We should be highly sceptical about talk of 'Revolutions in Military Affairs' (RMAs) based on narrow technological development. An RMA might now happen, if our improved paradigm allows us to see war, and address it, differently. Some people will disagree with this new view of war; that happens with paradigms. Nevertheless, the resulting debate should produce a far better understanding of the human face of war.

What I have just described is intended to be a journey. I use the term 'we' throughout this book in the hope that the reader will be drawn along a journey of exploration. I am to some extent a participant in the debate, so the occasional 'I' slips in as well. I was an infantry officer in an ancient and not particularly famous regiment of the British Army for almost 25 years. During that time I gained a doctorate for a study of the nature of military thought. I was the most prolific writer of military articles in the British Army for over a decade, which isn't saying very much. The comment, reflected in the foreword to this book, that I was 'about as popular as an Old Testament prophet' seems to have been intended as compliment, and got me promoted. For the last five or so years of my service I wrote military doctrine as a member of the General Staff. At one stage I was involved in a significant row between the army's doctrine branch and the Staff College. The row centred on some of the material in this book. I travelled quite widely and lectured in several countries. We held staff talks with a number of other armies, and exchanged ideas and perspectives. In my second career I still work in this area. I do some work at the British Defence Academy and undertake research for government and non-government bodies.

This is not a book of military history, although it inevitably contains some. It focuses mainly on the tactical level, so it is not about strategy or operational art either. Nor is it primarily about 'irregular warfare': insurgency, counter-insurgency or peace-support operations. Given my opening remarks, that should not be surprising. There is very little in this book about air operations. None of that should be taken to mean that, say, operational art, counter-insurgency or air–land operations aren't important. Of course they are. There seems to be a need for a similar book about irregular war. In the specific area of air–land operations, wherever I use the term 'land', as in 'land operations', I mean 'air–land operations'. But this book is unashamedly about regular land warfare at the tactical level.

I am grateful to many people for their help along the way. Some of them are civilians, who have generally given me ideas. Some have been soldiers, who have generally given me encouragement and opportunity. Some have done both. They include: Major-General Jonathan Bailey, Mike Bathe, Lieutenant-General Bob Baxter, Professor Karen Carr, Major-General Mike Charlton-Weedy, Colonel Mike Crawshaw, Dr Heidi Doughty, Bob Evans, Colonel Richard Iron, Lieutenant-General Sir John Kiszely, Major-General Mungo Melvin, Wilf Owen, Brigadier David Potts, John Storr, Joanne Suddaby-Smith, Brigadier Gage Williams and Colonel John Wilson. They also include many members of the former Defence Evaluation and Research Agency; not least Lorraine Dodd, Simon Henderson, the late Graham Mathieson, Professor Jim Mofffat, Dermot Rooney, Alan Robinson, David Rowland and Liz Wheatley. I also wish to thank Professor Gary Sheffield and Dr Dan Todman for their perceptive and useful comments on a draft of this book. For his unwavering patience, wise guidance and continuing support, I am extremely grateful to Professor Richard Holmes.

Any mistakes, of course, are mine.

1

Art or Science?

Bronowski wrote of Einstein that he asked some very simple questions, and that when the answers were simple too, 'then you hear God thinking'.[1] At the beginning of a book like this we should ask some very simple questions. What is military thought? What is it like? Who does it? The answers are perhaps surprising. There is relatively little military thought; it is done by very few people; and what exists is surprisingly incoherent and inconsistent. Furthermore, there seems to be no agreed philosophical approach to the subject. In practice we are concerned not so much with war as with waging it. That is, not war but warfare. But what is warfare, what it is like, and what we can say about it?

War is 'strife; usually between nations, conducted by force'; warfare is a 'state of war; campaigning; being engaged in war'.[2] Thus warfare is the conduct of war. We should differentiate between the study of war and the study of warfare. The study of war is largely an issue of history. Warfare, and hence military theory, is largely an issue of methods and methodologies. By analogy, medicine is 'a substance used in restoring or preserving health'; the practice of medicine is the 'art of restoring and preserving health'. There is a discipline, and an attendant profession, of medicine. In a way that is not particularly obvious, there appears to be a profession of arms and so one would expect a discipline of warfare. That discipline is the realm of military theory. Our task is to explore that discipline.

The profession of arms appears to be well defined. There are soldiers; their main purpose appears to be the conduct of strife, usually in formed bodies, and often representing nations. However, by the 'profession of arms', do we mean all soldiers, all salaried soldiers, or only officers, or only officers from combat arms? Yet that ambiguity is relatively small, compared with the ambiguity about the nature or even existence of a discipline of war.

We defined war as 'strife; usually between nations'. In this context, 'strife' is usually taken to mean collective armed violence. A nation is 'a large number of people of mainly common descent, language, history, etc., usually inhabiting a territory bounded by defined limits and forming a society under one government'.[3] Thus war reduces broadly to 'collective armed violence between states'. This is *not* categorical. The level of violence may be almost absolute (for example, the Second World War) or trivial (the Cold War). The definition of a state allows much to interpretation: when does a people or a society become a state? Were the

Palestinians a nation before the creation of the Palestinian Authority? Was the Intifada a war or an insurrection? Much has been written on this subject; perhaps the best approach is Sir John Keegan's suggestion of a national horizon.[4] When a people can identify itself as a nation, and organizes collective violence in pursuit of its goals, it might be said to wage war.

Even this is not categorical. To what extent does the Catholic community of Northern Ireland identify itself as separate from the remainder of the population? To what extent did Republican terrorist organizations wage collective violence on their behalf? To what extent was the conflict in Northern Ireland a civil war? Such questions appear to have no categoric answer. War does not seem amenable to close definition, thus neither is warfare. That does not mean that we cannot gain useful insight from its study. Unfortunately, since such questions do not appear to have categorical answers, the field is ripe for pedantry. The great danger of pedantry is that it undermines what is otherwise useful generalization. At this point, we are seeking insight. We should not expect categoric conclusions.

Our working definition of war is therefore 'collective armed violence between states', accepting that this is not categoric. Warfare must therefore be the conduct of armed violence between states, again avoiding categorical distinction. A discipline of warfare concerns the means and methods of conducting collective armed violence for the purposes of the state.

British military doctrine provides a useful taxonomy.[5] It divides fighting power into three components. The physical component is the means to fight: the equipment and the soldiers, and the logistics required to prosecute and sustain the fight. The moral component is the will to fight. It concerns morale, motivation, leadership and military culture. The conceptual component is roughly how to fight: the domain of thought. It concerns methods, approaches and the wider uses of thought on and off the battlefield. The term 'wider uses of thought' includes any other aspect of mental process. We shall explore the nature of military thought, hence the conceptual component of fighting power. It is not sensible to define it too narrowly, and possibly exclude fruitful areas of discovery, at this stage.

Military thought concerns the ability to *win* battles, engagements, campaigns and hence wars. States do not wage war for the sake of it; they wish to achieve certain objectives. Simplistically, achieving those goals constitutes victory from their perspective. Thus the value of military thought is largely to the practitioners. It helps them win wars. Apart from its use to by those practitioners, it would be of little interest other than to a small number of academics. However, since the fate of nations depends from time to time on the effectiveness of their armed forces, it is of some interest to much of the population. As we shall see, the subject is relatively poorly understood. To the practitioner its value lies largely in its contribution to fighting power. Its worth should be seen in terms of military effectiveness – the ability to *win* wars, campaigns, battles and engagements. Armies do not get paid to come second. This suggests an approach based on pragmatism: not primarily an issue of method, but of result. Methods are useful in so far as they support understanding, and enable positive results to be obtained. The value of studying warfare lies in its contribution to winning. Denigrating method in favour of result also tends to suggest an approach based on empiricism. We should find out what works, and do it; not dwell on theory for its own sake.

This pragmatic view may appear unduly workmanlike, but it is valid. A civil engineer must understand the laws of physics and the behaviour of materials, but his primary concern is to get things built. His main use of theory is as a tool. Other people, such as physicists and material scientists, are concerned with the development of theory for its own sake. However, the value of warfare as a discipline lies largely in its contribution to military success or effectiveness, which suggests an approach based on pragmatism and empiricism.

Perhaps the hallmark question in this area is whether the conduct of war is an art or a science. The question presupposes that there is a choice of only two alternatives. The question reflects an academic or intellectual paradigm now largely forgotten elsewhere. To a Victorian, the sciences were precise and in some ways useful. The arts were creative, inspirational and aesthetic. In this world view there was perhaps an assumption that sciences related to mechanical crafts and were vulgar and commercial.[6] The paradigm reflected the division of university studies, which resulted in degrees being awarded in either sciences or arts. However, academic and intellectual thought has moved on. Is medicine an art or a science? Neither – it is a discipline in its own right. Some aspects of medicine involve highly scientific issues. Other aspects, particularly where they relate to complex diagnoses, appear to be highly creative; although based on a solid basis of knowledge. Medicine can be seen to have 'artistic' aspects. Nevertheless, although medicine displays both artistic and scientific aspects, it is normally seen as a distinct discipline, with its own battery of techniques and approaches. Similarly engineering, economics and law.

By extension, warfare does not have to be an art or a science. A layman would consider some aspects to be entirely technical, and others to be creative – such as the use of bold manoeuvre on or off the battlefield. Other aspects are inspirational: the existence of leadership as an identifiable phenomenon requires it to be so. However, at its most basic, science is a matter of research. War and warfare are not. Soldiers do not conduct research when they fight. If they did, they would probably die. Warfare is neither art nor science. It should be considered a discipline in its own right. Clausewitz considered war to be part of man's social existence.[7] The prevalence of the question suggests that thinking in this area is atrophied. If the hallmark of discussion on warfare is an enquiry as to its nature in accordance with an outdated Victorian paradigm, then the discipline lacks vigour; let alone rigour.

When we look at warfare, we find a poorly developed discipline. It is incoherent, contains a range of poorly described phenomena and is pervaded by paradox. There are few classic texts. Clausewitz's *On War* is certainly one of them, but very few people today have read it. It is written in early nineteenth-century German. For most English speakers it is very difficult in translation and impossible in the original. Apparently German Army officers prefer to read the English translation! The two translations I have looked at vary appreciably. *On War* has had considerable impact on Western military thought, but since few British or Americans have read it at all, let alone in the original, one must wonder whether they truly understand what Clausewitz meant. As the Israeli historian Azar Gat pointed out, the text we have today is confusing and incomplete because Clausewitz was in the process of revising it when he died.[8] There is also a tendency to cite Clausewitz highly selectively. Besides, does it really matter precisely what Clausewitz meant, 200 years ago? Surely it is more important to develop a better concept of war broadly grounded in his ideas.

If much of this criticism applies to a comparatively recent writer, how much more so of the ancient Chinese writer Sun Tzu? Yet these criticisms do not even touch the content; just the nature of the texts themselves. More recent military theory can be found in two broad areas. The first area is conceptual writings by authors such as the British Captain Sir Basil Liddell Hart and Major-General J.F.C. Fuller. Later examples include the Israeli historian Professor Martin van Creveld, and the American Dr Edward Luttwak. The second area is military doctrine.

In looking at these works we search in vain for an underlying theory of war. Clausewitz undeniably lays out some useful tenets, such as that war is the extension of politics by other means. Fuller did a great service in enunciating a list of Principles of War. Soldiers tend to think that Principles of War are all that is needed as a comprehensive military theory. However, they are not quite as cast in stone as they might think. A list of eight principles was developed by Fuller in 1916.[9] The list was then amended roughly once a decade until 1968. The number fluctuated, as did the content of the list. Field Marshal Montgomery changed the list considerably, but his changes scarcely survived his tenure as Chief of General Staff. The list we have today is broadly Fuller's final list of ten. The US army's current list is nearer to Fuller's original version than is the British. The Principles of War are by no means immutable, and other nations do not use the same list.[10]

Nevertheless, they are a standardized list of what the British or American armies believe to be important. Perhaps, unsurprisingly, that has changed over time. In the long run, military doctrine may be no more than an army's current best guess as to how to fight at the start of the next war.[11] The Principles of War serve that purpose moderately well, but they don't answer the question: how should armies fight? What should units and formations do in battle, in order to gain tactical success and then win campaigns and wars? At best, the Principles of War give broad guidance as to the sorts of things that are likely to be useful. They provide no concept of mechanism; no description of what happens in battle; how one side wins and the other loses.

Furthermore, few people are engaged in developing military thought. No one has ever made a full career in this area. This compares poorly with other disciplines. In 2001 there were 97,436 doctors in the National Health Service in Britain,[12] and some thousands more working in private practice. Several thousand worked full time, or for a significant part of their time, in research. University civil engineering departments produced perhaps 600–1,000 graduates per year. Those departments typically employed 30–40 academic staff each.[13] By contrast, the British Army's doctrine staff numbers fewer than 20. A number of officers write, but relatively few write about warfare, and very few ever write more than one article. Only a handful of civilian authors write on warfare.

The academic caucus is tiny: three or four university departments. Most if not all of their staff are International Relations (IR) experts or historians. The number actually writing about warfare is tiny. In my experience, IR experts discussing warfare tend to overgeneralize from singular or few facts; discuss the discourse rather than the events; use other academics as authorities; use terms outside common usage (such as 'postmodernism' or 'the entrepreneurial use of the politics of identity'); and make remarks that are blatantly counterfactual, such as 'the end of war'. They also tend to use history non-rigorously, cherry-picking historical examples.

Respected historians write perhaps one major work on warfare each. Most historians do not appear to be particularly numerate, which is a significant problem. Even within published military history we find significant gaps. I know of no convincing account of how, for example, the campaign in north-west Europe in 1944–45 was fought and won. Similarly, more than 60 years after the end of the Second World War, there is no clear consensus as to the effectiveness of the Allied strategic bombing campaign. But surely those were the most significant campaigns waged by the Western Allies in that war? No one appears to specialize in warfare as a discrete discipline.

Furthermore, the body of knowledge about warfare is small. The library of the Royal Military Academy at Sandhurst, or that of the Joint Services Command and Staff College, is no larger than that of a single large university department. The sections on warfare extend to a few books. Bearing all this in mind, it is scarcely surprising that warfare appears incoherent, poorly described and pervaded by paradox.

Around the year 2000 military writers began to discuss something variously called 'effects-based operations' (EBO), 'effects-based approaches', and similar. The progress of that discussion gives a fascinating insight into current military thought. Suffice it to say that by 2005 no real consensus as to what it was, let alone how to do it or whether it was worth doing, had emerged. Yet in 2002 a rather suave senior officer at the British infantry school had announced with tremendous self-confidence that 'everything here is now effects-based'. At a subsequent briefing to a German army delegation, a presenter announced with great flourish that, essentially, whenever one considers a course of action one should think of its effect on the enemy (and possibly also the environment and one's own forces). The German visitors were of course polite in response, but scarcely impressed!

Colin Gray considered EBO to be both unmistakably banal and dangerously illusory.[14] One finds in the effects-based discussion an emotive, superficial and often politically motivated line of argument. For example, much of the original thinking in the area appears to have been prompted to justify the budget for US stealth bombers. Buzzwords, jargon and poorly defined terms abound, and terms are employed with the flimsiest justification.[15] If we ask simply, 'how do we know that an effects-based approach is right', or 'do we actually know that it is better than what we used to do? And if so how?', there appears to be no good answer.[16] I am continually disappointed that many military men do not detect the flimsy intellectual content of such constructs.

The use of the term 'kinetic' is a particularly bad example of poor military thought. It is used to described violent effects, typically in the context of 'effects-based operations' and similar. In every case I have found, the terms 'violent' and 'non-violent' would be more accurate than 'kinetic' or 'non-kinetic'. For example, only the pedantic or the lazy can use the term 'kinetic' to describe the effect of a bomb going off. It seems that 'kinetic' is preferred because it seems somehow more sophisticated. It is not. It is just sloppy use of English. There does appear to be something useful in effects-based approaches, but the discussion surrounding the subject has been dreadful. If that is the best we can do, it is no wonder that military thought is poorly developed.

In the early and mid 1990s much military thinking revolved around the so-called 'OODA Loop'.[17] The OODA Loop suggests that the basic process of command and control (C2), described as Observation, Orientation, Decision and Action, is a circular and iterative process. Military advantage is supposed to accrue from being able to go around the loop faster than one's opponent.

However, the C2 process is not circular. According to the training pamphlets of the time, it apparently took 24 hours for a division to execute a divisional-level operation.[18] Planning took a further 12 hours at least. Thus a divisional OODA loop would have to be at least 36 hours long. Allowing a reasonable time for action to take place and be observed suggests an even longer loop. Yet the Gulf War of 1991 and the fighting phase of the Iraq War in 2003 showed divisions reacting far faster. Commanders and staffs do not in practice wait to observe until after they have acted. They observe continuously, and act when required. The relevant action is not the action of the commander and his staff, but that of subordinates at some distance in time and space. The headquarters' (HQ) action is largely limited to preparing and giving orders. In reality Observation, Orientation and Action are continuous processes. Action is continuous in the sense that, from the HQ's perspective, some action is taking place which it can observe. Decisions are made occasionally as a result of those actions.

The key fallacy in the OODA Loop is that the C2 process is not iterative in the accepted sense. The idea of 'getting inside the enemy's decision cycle' is deeply flawed; not least because every circuit of the loop must involve some interaction of forces and hence attrition, which Western doctrine avoids where possible. Armed forces should not choose a strategy based on iteration. There is considerable advantage in reacting faster than one's opponent,[19] but the OODA Loop does not adequately describe the process. It places undue emphasis on iteration, as opposed to tactically decisive action.

The original concept was based on fighter combat in Korea. The American writer Bill Lind based his ideas about OODA on USAF Colonel John Boyd's observations about F86 Sabre Jets and their pilots. Lind extrapolated from there to command and control in general. This is an exercise in induction (the generation of general statements from particular ones). Induction is philosophically unsafe.[20] To generalize about formation-level C2 from aircraft design is tenuous, to say the least.

The epitome of fighter combat is the performance of the tiny cadre of ace pilots. An 'ace' is a fighter-pilot who has shot down five or more enemy aircraft. Historically, aces constitute a very small proportion of all pilots, but account for a very high proportion of all aircraft shot down. For example, half of the 25,000 Western aircraft shot down by the Luftwaffe were shot down by fewer than 500 pilots, from a total of at least 20,000.[21] Unfortunately, biographies of aces such as General Robin Olds, who shot down at least five aircraft in each of three wars – the Second World War, Korea and Vietnam – show almost no trace of iterative behaviour in combat.[22]

Critically, aces scarcely ever dogfight. They usually destroy enemy aircraft with a single pass, and expend very little ammunition per aircraft shot down. Their effectiveness centres on rapid, decisive decision and action. It is based on superlative, largely intuitive, situational awareness. Aces do display some significant characteristics – their eyesight is usually exceptional and their shooting phenomenal. They also have catlike reactions.[23] However, expert fighter combat is fundamentally not iterative. It is sudden, dramatic and decisive. Thus

Lind's concept of an OODA Loop does not adequately describe the observed facts for the activity under study – fighter combat – let alone any extrapolation from them.

This excursion into fighter tactics is useful for several reasons. First, it exposes a major weakness in modern military writing: the OODA Loop is flawed. Secondly, it demonstrates the importance of examining the observed facts, and not the accepted theory. Thirdly, by extension, it raises a question of authority. The OODA Loop was incorporated into British military doctrine in the 1990s.[24] Galileo Galilei, writing in seventeenth-century Italy, was forced to choose between Catholic Church doctrine (which stated as an article of faith that the Earth does not move; the Sun revolves around the Earth) and his own observations, which told him otherwise. One of Galileo's most important contributions to knowledge is that in the choice between an appeal to authority and the observed facts, one should believe the facts.[25]

The OODA Loop is an example which originated in conceptual military writing. Turning to doctrine, the British military doctrine of the 1990s was underpinned by two sets of functions: the Core Functions (Find, Fix and Strike); and the Functions in Combat (Command, Manoeuvre, Firepower, Information and Intelligence, Protection and Combat Service Support). These were stated authoritatively. The reader was invited to accept them as an article of faith without justification.[26] There was no established connection between the two sets of functions, identified in the key document in unconnected chapters. And there was no concrete linkage between the various functions and the Principles of War.[27]

As a further example, consider the term 'Command and Control'. 'Command' is defined as 'the direction, coordination and control of military forces'.[28] 'Control' is 'the process by which a commander . . . organizes, directs and coordinates the activities of the forces allocated to him . . .'[29] This is clearly nonsense. Those two definitions are circular, which makes the expression 'Command and Control' fairly meaningless.

Much doctrine appears incoherent and shows evidence of having been written by committees.[30] That same British doctrine stressed shock and surprise, but contained no definition of either, let alone a description of how to inflict them. The concept of 'the initiative', much discussed within the army, was not defined, and merited a single paragraph in the formation tactics handbook.[31] It was surprising to find that published doctrine is not particularly coherent. This is not just a criticism of British doctrine. Similar comments were made on US, Canadian and NATO doctrine in the mid 1990s.[32]

The US Army's capstone publication *Field Manual (FM) 100–5* of 1994 was another good example. It provided lists of principles, tenets and functions.[33] Each group was moderately consistent in itself, but provided virtually no link to the other. How do the 'enduring tenets' link to 'functions in combat'? Where do the

Principles of War spring from, and how do they relate to tenets and principles? What is their basis in fundamental behaviour? As with the British ADP series, high-level statements are made authoritatively, with no supporting references. In general, much doctrine gives the impression of dogma, or of 'slabs of thought held together by their own self-contradiction'.[34]

Major elements of military thought are demonstrably missing. Napoleon's operational design revolved around the '*bataillon carrée*'. Prussian nineteenth-century operational tenets revolved around the formula of 'Reconnaissance, Victory and Pursuit', and the technique of double encirclement.[35] Yet we see nothing equivalent in modern British and American thought. The best we see is a methodology for *developing* campaign plans.[36]

In the mid 1990s British military discussion embraced the term 'the Rule of Four' as a 'structural principle'. Put simply, at each level a tactical commander required four principle subordinates in order to give appropriate flexibility on the battlefield. So, for example, a brigade should contain four battalion-sized units. Under the Rule of Four, a commander needs one subordinate to fix the enemy, another to strike, a third as an echelon force to sustain the tempo of the advance and a fourth as a true reserve to be held against unexpected contingencies.

The term was used loosely in discussion until the staff of the army's doctrine branch drafted a discussion note. The Rule had become a Principle, since principles allegedly admit of exceptions while rules do not.[37,38] The 'Four' had become 'Three Plus', on the realization that one of the subordinates could be an artillery unit. The Principle was not required to apply at sub-unit level, on the grounds that companies had no need for a true reserve. However, the 'Three Plus' concept would lead automatically to attacking 'one up' in normal circumstances (that is, one to strike, one as an echelon force and one as a true reserve), which was demonstrably wrong. In addition, a 'principle of four' is inherently inefficient. A brigade of 16 companies in four battalions would only place four in the first echelon. Unfortunately, a brigade of nine companies in three battalions would also place four companies in its first echelon. Therefore in the first instance the larger brigade is using seven more companies to achieve the same effect. (We shall return to this issue.) And finally, the best available evidence of the use of the Rule of Four at company level and below suggests considerable advantages.[39] The evidence was not conclusive, but contradicted the idea that the Principle of Four is not needed at sub-unit level.

Some years later I found the original staff papers relating to the concept. All of them seemed to be based on assertion. There was no convincing evidence to support the Rule or Principle of Four. The historical examples cited were flawed. It seemed to be a major case of what people wanted to believe, and may have been largely the views of a single influential person. The Principle of Four reveals a lack of rigour, and shows signs of being forced to fit the situation. The history of the Principle of Four demonstrates many human aspects of military thought rather than any great theory. It was a myth: something with a kernel of truth, but which was widely believed to be true. It was only in August 2000 that any methodical study was initiated to investigate the principle. That was after it had been taught as endorsed doctrine at Staff College and used to support the size and shape of the army during the Strategic Defence Review of 1997–98. The manner in which it arose, was codified, applied and then justified after the event says much about military thought.

In practice, much of what we see in military thought seems mythological. It has little base in fact and contains many half-truths and legends. There is an almost pathetic grasp of neologisms such as 'EBO', 'network-centric' or a 'Revolution in Military Affairs'. Like the worst aspects of religion, it contains elements of what people want to believe and things that they find politically acceptable or useful to believe, using the term 'politics' in its widest sense. Ideas emerge, bubble up, and may or may not be embraced in formal teaching. They may then disappear 'when interest wanes, circumstances alter, or the next big idea comes along'.[40] What is formalized is often incoherent and inconsistent. In short, there are fads and fashions in military thought.

This is not just a shame; it is a problem. The existence of a coherent body of thought can have a devastating impact on military effectiveness. Relative to their enemies, the armies of Revolutionary and Napoleonic France had such a body of thought. With that, large armies, and a generous dose of revolutionary zeal, they swept across much of Europe for a couple of decades. Or consider the German Wehrmacht at the beginning of the Second World War. Relative to their enemies, they were much better prepared doctrinally, and that gave them one of several advantages on the battlefield. But note: 'relative to their enemies'. Neither French Revolutionary nor Wehrmacht military thought was perfect, but both were much better than that of their enemies. So, although Western armies can and should do better, all is not bad. Their doctrine is much better than that of most of the people they are likely to meet. It is also generally better than it was, say, 20 years ago.

Military writing contains a lot of paradox. Clausewitz dwells on paradox: because of it, war is unlike the rational relations of science or the creative aspects of the arts.[41] Edward Luttwak makes much of it in *The Logic of War and Peace*.[42] There are many examples of paradox in military writing. 'Attack is the best form of defence' is one. 'If you wish peace, prepare for war' is another. The suggestion that peace could be maintained between superpowers by the possession of large nuclear arsenals is a third. In the advance you should choose the worst road; it is actually the best, because the enemy will not expect you there. A defence can be too successful: it may consume more assets than it merits. Geometrically, large countries have small perimeters relative to their land area, but greater resources and population. Therefore they should be easier to defend than small countries. However, large countries evoke fear in their neighbours, and therefore may be attacked more frequently.[43] These are all examples of paradox, found in theoretical military writings from Vegetius to Luttwak. More recently, the new US manual on counterinsurgency has a section which describes nine separate paradoxes. The section on logistics describes two more.[44]

A paradox is a pair of statements which appear to be true but contradict each other.[45] Paradoxes are often presented as single statements, such as 'attack is the best form of defence'. Strictly, however, such a paradox comprises two elements. In our example, one might be 'the attack is an aggressive act which may result in high casualties to the attacker, whilst the defence is primarily intended to protect one's own forces'. The other would be 'attacking often has the effect of causing less harm to one's own forces than defending', or similar. As in this example, one element (the first) is often implicit. Expanded in this way, the contradiction between the two elements becomes clearer.

The resolution of a paradox requires our underlying view of the situation, our paradigm, to change in some way.[46] In this manner, either one statement no longer appears true or, more commonly, the two statements no longer appear contradictory.[47] The classic paradox is the Paradox of Zeno, often known as 'Achilles and the Tortoise'. The Ancients never resolved Zeno's Paradox. In practice, they never developed a satisfactory understanding of dynamics. Today the resolution of Zeno's paradox is straightforward. It requires a knowledge of mathematics (particularly the summation of series, and elementary differentiation) which was only developed in the seventeenth century. In other words, it required a fundamental view of the world, and discrete mathematical tools, which were only developed many centuries after Zeno's death.

Man's view of his world is rarely perfect, so paradoxes do occur. However, a prevalence of paradox should tell us that our understanding of war, and warfare, requires fundamental adjustment. This is a major concern. When esteemed authors write whole books on paradox in war, alarm bells should ring long and loud. We should have no confidence in 'paradoxical logic', as espoused by Luttwak and even Clausewitz.[48] Paradoxical logic is not logic at all, but a strong indicator of misunderstanding.[49] Fundamentally, the identification of a prevalence of paradox tells us that our view of war, our underlying paradigm, is deeply and fundamentally flawed. We should explore the underlying paradigm, resolve its inconsistencies, and thus come to a better understanding of war and warfare. Indeed, it may be that the military predilection for fad, buzzwords and poorly enunciated jargon reflects the absence of a unifying paradigm of war.[50]

Such a fundamental intellectual approach or philosophy is clearly missing from British and American works on warfare. At the very least, if there is such an underpinning approach, it is not explicit. Without that common understanding or paradigm, paradox will continue to flourish. The absence of an explicit and consistent mental approach makes it very difficult to develop a discipline of warfare.

The American philosopher Professor Thomas Kuhn's landmark book *The Nature of Scientific Revolution* is grounded in the concept that the development of a science – such as, for example, botany, theoretical physics or microbiology – depends on a shared view of the subject and a shared approach amongst its practitioners.[51] The shared view extends to an understanding of which phenomena are within the scope of the discipline, what problems are significant, and which approaches are likely to be fruitful in solving them. It tends to result in a common terminology, which is surprisingly lacking from military thought – at least in Britain, if not the English-speaking world. Until such a shared approach emerges, it is very hard (if not impossible) for an intellectual discipline to progress.[52] For example, until a common view of molecular theory was developed by Joseph Priestley and others in the eighteenth century, it was almost impossible for chemistry to develop.[53]

Current approaches to the conduct of war are intellectually shallow. General Fuller seems to have been the first to enunciate lists of principles, taken forward as key words or phrases.[54] Lists of principles were developed by various twentieth-century war leaders such as Montgomery and Slim in their 'Qualities of a Leader'. Such lists lack overarching theory, and provide little real insight. The principles they contain appear to be largely co-equal; are of unknown origin (hence validity); and are obviously culturally related (that is, they necessarily reflect the values that their author wishes to share with his audience). They are therefore of limited application. They consequently tend to trite anecdotal use.

Should we expect more? In any established discipline, the great body of actual experience is typically remote from the practitioner. The average fifth-former has not actually experienced most of the science he describes in his GCSE exam papers.[55] However, the science has been conducted; more advanced treatises do describe it; and it is accessible to fully qualified practitioners if required. Conversely students at Staff Colleges tend to find nothing more profound than the pamphlets we have already criticized. So yes we should expect more.

In order to understand why military thought is relatively undeveloped, we should look at its methods. We have seen that the underlying paradigm used in British military thought is flawed. Yet even with a flawed paradigm, one can make useful progress – man built Gothic cathedrals before Newton developed the understanding and the mathematics required to analyse the structures involved. So, how do people approach military thought?

Most civilian writers approach warfare from military history. Their observations are based on historical fact, and they make deductions – often most perceptive – based on those observations. However, they do not attempt any mathematical treatment of such history and certainly no statistical analysis. There is little constructive theory and, where there is,[56] other historians appear to denigrate it.

Military writers normally use a narrower and more explicit methodology. British military writing is encapsulated in the process of 'Service Writing'. It includes minor typographic conventions, and more fundamental conventions regarding content. Substantive works follow the conventions of a 'Staff Paper', whose structure and layout are formalized in military regulations. The overall process is to describe an issue, state the aim of the paper, consider a number of factors and draw deductions. The paper is then summarized, deductions are listed and recommendations made. The recommendations should flow naturally from the deductions. The factors are to be considered in a 'balanced' manner, which broadly equates to presenting both, or all, sides of an argument. Overt reasoning – explicit logic – is prized, and there should be a logical flow through the paper without obvious gaps or jumps.

This is a good description of a rationalistic method. Rationalism derives from the (predominantly French) Rationalists, typified by Descartes and Pascal. Rationalistic methods seek to gain knowledge through the exercise of reason alone.[57] Such methods are extremely powerful: Descartes, Pascal and Leibniz are the fathers of most modern mathematics; Leibniz was perhaps the most universal genius of modern times.[58]

Rationalism is essentially deductive: statements of the particular are derived from statements of the general case. This is philosophically safe, but has limitations. In principle it is possible to continue to make further and further deductions and thereby construct detailed theoretical frameworks. In practice such methods fail, because inherently they make little reference to facts outside the initial general conditions – the statements of the general case.

We can take an example from geometry. We observe that there are things called parallel lines which never meet. From there we can make deductions about the angles contained in the interstices between a pair of parallel lines and a third line that crosses both of them. By adding another line we can make further deductions about the internal angles of a triangle: they must add up precisely to 180°. This was known by the Ancient Greeks. The resulting theories of geometry are known as 'Euclidean'. They all revolve around the general statement that the internal angles of a triangle add up to 180°. However, it is easy to describe a triangle whose angles do not add up to 180°. Simply draw the triangle on the surface of a sphere.

The underlying reason is that our initial observations about parallel lines that never meet are only true on a flat surface. This is obviously a special case, although a highly important one. The field of geometry which stems from triangles whose angles do not add up to 180° is important, not least in cartography. It is non-Euclidean geometry. In practice we have gone beyond the initial statement of the general case. Instead of 'there are things called parallel lines which never meet' we should have said 'there are things called parallel lines which never meet *on a flat surface*'. In general, rationalism has considerable limitations. The further we move from the immediate, observed facts using rationalistic methods, the less confidence we can have in our findings. Thus strong rationalism can sometimes serve us badly in military thought.

This begs the question of what are the facts, and how far we are from them. Rugged military men, writing on things familiar to them, are unlikely to go badly wrong. However, as first-hand experience of old-fashioned war diminishes, and novel technology and a changing world scene render old truisms irrelevant, rationalism is likely to be found wanting. Arguably Western armed forces are at that stage in the first decade of the twenty-first century.

A collection of incoherent statements, poorly described phenomena and paradox is typical of the early stages of the development of an intellectual discipline.[59] They can be likened to early natural histories, such as Gilbert White's landmark *Natural History of Selborne*, written in 1778–9.[60] Such works can appear erudite and well written, not least because they describe the current state of knowledge at the time of writing. Unfortunately, while being thoroughly descriptive, they fall well short of being prescriptive. They list what can be observed, but can give no reason why it happens, or how to make it happen.

Natural histories also often omit details subsequently shown to be extremely revealing. Conversely they tend to include irrelevancies, because the writer has no conceptual framework by which to know what kinds of phenomena are important in the field of study.[61] A good example of such irrelevance comes from Newton's study of thermodynamics (the transfer of heat). Newton correctly observed that combustion was a source of heat, and also that heat is a property which can be transferred. At the time, this was ground-breaking development. However, Newton also observed that dunghills were sources of heat. This is true but irrelevant: it results from the anaerobic, chemical decomposition of organic matter.[62] They don't burn: they rot. It is comical today but remained a serious discrepancy for decades after Newton.

The underlying tenet of Kuhn's work is that the history of science is dominated not by objective facts but by the sociology of science. The development of a discipline reflects many human issues, such as which problems are important. For example, astronomy was critically important in Western Europe from the fifteenth to the nineteenth centuries because of its relevance to maritime navigation. Intellectual disciplines have a strong sociological underpinning. What can we say about the people who write about warfare?

The two communities are either military officers, serving a two- to three-year tour of duty in a doctrine centre; or academics with a background in history or international relations. That will tend to influence the way those two communities view the discipline, and therefore the work they produce. It may be suggested that, contrary to their protestations, military doctrine writers are primarily concerned with producing instructional pamphlets. A second activity is to support policy decisions made elsewhere. Many seem to be better at mastering the relevant bureaucratic processes than conceptual thought. In my experience, fundamental thinking is rare. Much material is recycled for ease, and in a quest for consistency. Genuinely novel insights can easily be sidelined due to the issues of the moment.[63] The academic community treats theoretical writing as an extension of history, to be written after a relatively long career as an academic historian. Neither community is likely to achieve a profound understanding of the basic phenomena.

Therefore it appears that we have a discipline which, surprisingly, shows signs of being in the early stages of development. It is enunciated by two communities who are systematically unlikely to develop a deep understanding of the underlying phenomena. The picture is not encouraging.

2

Developing an Approach

We should now consider which mental or philosophical approaches to use when studying warfare. One view of mental endeavour would consider all intellectual disciplines spread along a single axis. They would be located according to their apparent deterministic content. Events are described as 'determined' if they behave according to observable laws. Intellectual disciplines would be placed on our axis according to the extent to which the phenomena with which they deal obey observable laws. The 'hard' sciences such as physics, chemistry and biology would lie at one end. Human behaviour is not generally yet thought to obey observable laws, and so 'softer' sciences would lie around the middle. Towards the other end would be disciplines which appeal directly to aesthetics or faith, and hence have little apparent deterministic content. Art or religion would belong there.

So, using our Victorian paradigm, we have a range from Science to Art. Let us now consider medicine and engineering. Neither is 'art' nor 'science': it is perhaps belittling to describe them as either. Much of medicine is well described, and shows little paradox or inconsistency. In engineering explicit theory sometimes appears at the design stage. However, at the point of construction or production issues such as site organization, workforce motivation and even customer preference come to the fore. Medicine and engineering have been practised in some recognizable form for millennia. Both are established and mature intellectual disciplines. One might reasonably expect the same of warfare, yet we have seen that is not the case. The underlying reason appears to lie in the nature of determinism.

Determinism requires events to occur according to fundamental laws. In principle such events should be predictable, even if the underlying laws have not yet been discovered. There is overwhelming evidence that the universe is in fact determined.[1] However, war does not seem to be like that: it does not *appear* to be determined. 'Every science has principles and rules', wrote the Maréchal de Saxe; 'only war has none.'[2] Clausewitz agreed.[3] Furthermore, one of the criticisms of Jomini is that he tried to suggest fundamental laws of warfare.[4] As John Keegan put it, 'Nothing in human affairs is predestinable, least of all in an exchange of energy as fluid and dynamic as a battle.'[5]

Determinism requires causality and continuity. If A occurs *and then* B occurs, we infer that A causes B. Note that we cannot observe causality, only infer it.[6] If A causes B *everywhere and at all times*, we consider it to be a principle of nature.[7] Such continuity requires that 'that which is a property of nature at any one time and in any one place, constantly and everywhere recurs'.[8] For all we know, war may be determined. It might eventually be possible to predict human behaviour from biochemical processes. To do so would require immense computing power, and knowledge of the precise state of every participant down to the molecular level at the start of the battle. However, that would be extremely difficult to measure, and may well be unknowable.[9] It is unlikely that even the most powerful computer will be able to achieve the necessary precision in the foreseeable future.[10] Thus the likelihood of being able accurately to predict the outcome of a small engagement, let alone a war, is vanishingly small. To that extent *war may be determined, but it is not useful to consider it so.* This is hugely important. It is absolutely fundamental to any intellectual discipline of warfare.

After centuries of thought, we have failed to find any strict relation between cause and effect on the battlefield. Continuity is weak and causality is almost nonexistent. This is our greatest challenge. It is probably the main reason why warfare is such a poorly established discipline. Since the fourth century, or even earlier, Western man has largely progressed by exploiting determined phenomena.[11] Yet in war he cannot. This is a pity: one observes a precision of expression in the philosophy of hard, determinist science which is often missing elsewhere. The philosophical works of Karl Popper or Ernst Mach are examples.

If we scratch the surface of military thought, and particularly British military thought, we see the insidious effects of determinism. We also see a mechanical paradigm. Britons tend to see the world in terms of cause and effect, input and output. This seems to have infected their view of war. The effect should not be understated. It tends to deny the British the opportunity to see the underlying nature of war. It is no surprise that Clausewitz was German. Similarly it is no surprise that a British officer, Ernest Swinton, predicted the need for a machine like a tank as early as 19 October 1914,[12] months before the true nature of trench warfare became apparent.[13]

In some way that we have not yet discussed, war is essentially a human activity. Some of the human behavioural sciences are, at least, well described – for example, sociology, psychology and anthropology. These developments have all occurred relatively recently – in the last 40 years or so.[14] A great deal of work has been done in understanding war in its wider (and especially its social) context, but this has not reached into the area of military theory.[15] The behavioural human sciences cannot yet predict human behaviour in any detail, and they also do not yet seem to have penetrated the thought processes of military thinkers.

That is perhaps surprising, given the impact of John Keegan's seminal work, *The Face of Battle*, in 1976.[16] *The Face of Battle* transformed much of military history by moving away from 'battles and generals' to a study of the soldiers themselves. Arguably it has had much less impact on warfare. This may say much about the sociology of doctrine writers. The effect may be generation-dependent. I was 16 years old when *The Face of Battle* was published. It conditioned much of what I subsequently read and observed. An older student might not have been so impressed by its appearance. It may be only now (as the current generation of doctrine writers are old enough to have read *The Face of Battle* as young men) that human issues will come to the fore.

In general, the behaviour of large numbers of human beings does not currently appear to behave in accordance with deterministic rules. If war were determined, we could make strong deductions from observed facts. Cause X would lead reliably to Effect Y, and so rationalistic methods would be useful. War does not seem to be strongly determined; least of all in its human aspects. To understand cause and effect, one needs to know the facts. Unfortunately facts are typically scarce, and often will be after protracted periods of peace. Therefore relying on rationalistic methods is doubly flawed.

If war is not determined, rationalism fails us. So also does the classical application of scientific method. Scientific method depends on observing facts, generating hypotheses, validating those hypotheses through experiment and subsequently observing new facts.[17] To be useful, experimental results must be repeatable: a given set of conditions (causes) should always lead to one and only one set of results (effects). This does not happen in war. Thus scientific methods fail us. In passing, it also strongly undermines much of the discussion of 'effects-based operations'.

Modern scientific method derives from the largely British (and generally English) school of Empiricists: Bacon, Locke, Berkeley, Hume and Mill. The fact that many were English is not coincidental – their contributions should be seen in the context of a nation busily attempting to expand both its overseas territories and its manufacturing capabilities. Empiricism had a devastatingly positive impact on human understanding, since it formalized the methodology of man's exploration of the physical world.[18] Thus, from the eighteenth century, man's quest for knowledge has largely been expressed through scientific techniques: an apparently endless quest for greater certainty and truth. We now see that that approach is inappropriate in warfare.

However, the fundamentals of empiricism are more encouraging. Not least, empirical enquiry can be lethal to a popular concept or fad.[19] The Empiricists' quest was not for ultimate truth, an accusation they raised at the Rationalists. Empiricism was intended only to develop probable hypotheses about the world around us.[20] This seems a reasonable approach to warfare, and suggests a way ahead. Our approach should be based on a little empiricism together with a generous dose of pragmatism.

If warfare is a poorly developed discipline, we should embark on a quest for more or better knowledge. Fundamentally, human knowledge can only proceed in two ways: by gathering new facts, or better organizing those which are already available.[21] Gathering relevant new facts is almost impossible for soldiers outside of major wars. Little of the 'savage wars of peace' resembles major war sufficiently well to be confident of their lessons. In the absence of new facts, armies seek to draw more and more understanding from a declining base of knowledge. The base declines because, outside major wars, we collectively forget much of war's complex detail. And the better organization of existing knowledge is, currently, largely conducted through those rationalistic processes which we have already rejected.

The major question is therefore: where to look for knowledge? There are in essence two options: observation and authority. Galileo made two landmark contributions to human knowledge. His specific, factual contribution was that the Earth is not fixed, but moves around the sun. As we have seen, his general, philosophical contribution was that in the choice between observation and authority, one should cling to the observed facts. The authority he referred to was Church doctrine, which was entirely contrary to his observations.[22] A military writer has a similar dilemma. He is to some extent constrained by extant military doctrine. Military doctrine is authoritative by intent. Yet we have already seen that much military doctrine is incoherent, inconsistent and poorly developed. Galileo's view is quite clear: observe the real world, and make deductions from there. As an aside, we should quite rightly be sceptical of military doctrine.

Yet, in the relatively long gap between major wars, there is relatively little to observe which may be relevant to future conflicts. There is a much richer seam of fact in military history. Naturally, since war is not determined, it will never repeat itself. Even identical conditions would never bring about the identical results. Yet, given a deep knowledge of military history, it should be possible to construct some hypotheses about future war which are good enough to serve an army until it has an opportunity to learn during its next conflict.

Lists of principles (such as Fuller's *Principles of War*) are an attempt to do just that. They are a limited attempt, and we should be capable of better. Empiricism can be used in a limited sense: observe the external world, not least through the deep study of military history; make deductions; and form hypotheses. The scientific method then requires those hypotheses to be tested through experiment. That will fail. Our experiments will not be conducted in realistic conditions, not least because we do not actually shoot people in peace. In addition, no such experiment would be truly repeatable.

Critically, however, such experiments give insight. If war is inherently complex (a theme which we will develop later), then such experiments give insight into the interactions that can occur in such complex circumstances. The results will never be more than generalizations. This is critical. The best that such experiments can provide is general insight into 'the sort of thing' that may happen in war. Scientists will, and do, complain that the results are not valid because they are not repeatable. Soldiers will, and do, complain because the experiment will not give 'the answer'.[23] There probably is no single answer, and we must accept this.

This is a pervading problem. It is not limited to warfare: it applies to science as a whole. Most of science is in practice an attempt to construct general theories from particular facts. That is an exercise in induction (the process of moving from the specific to the general). As Karl Popper, the philosopher of science, pointed out, induction is fundamentally unsafe. However, Popper continued, '[w]e do not know; we can only guess'. Such guesses are guided in practice by the metaphysical, the unscientific and by faith. 'But these marvellously imaginative and bold conjectures or "anticipations" of ours are carefully and soberly control-led by systematic tests.'[24]

We should not try to prove our hypotheses; we should try to disprove them. Francis Bacon pointed out the superior power of the negative instance: put simply, that it is generally difficult to prove something, but easier to disprove it. Popper deduced that in general no theory can be proven to be true. One can attempt to disprove a hypothesis, and if it passes a number of reasonable attempts at falsification, it can be taken as a valid theory for the time being. To that extent every hypothesis remains tentative for ever. It may be corroborated, but may subsequently be falsified. It is not the possession of knowledge which makes man the scientist, wrote Popper, but his 'persistent and recklessly critical *quest* for truth'.[25]

Popper's original discipline was theoretical physics: a vastly more determined field than warfare. His philosophy is, nevertheless, useful. Make conjectures – hypotheses beyond the scope of what you can observe – and seek to falsify them. If they stand up to attempted falsification, adopt them as useful theories for the time being. But keep observing the external world; keep seeking to falsify what we believe to be true; and move forward by creating more useful hypotheses. This is an intellectual discipline based on limited empiricism.

This discussion of 'hypothesis' and 'validation' may sound very wishy-washy in connection with military doctrine. However, it reflects a perceptive and pragmatic remark about the true nature of doctrine. In the final analysis doctrine is only the 'organization's current best guess about the best way to fight a war'.[26]

War is not for waging but for winning. Armies do not get paid to come second, not least due to the severe penalties incurred in losing. Useful military theories relate to winning. We want things that work; not merely things that are elegant or intellectually pleasing. This is pragmatism: an appeal to find 'what works'. Pragmatism developed from the work of William James into a sophisticated philosophy which is primarily a method of solving or evaluating intellectual problems. It is, however, also a theory about the kinds of knowledge we are capable of acquiring. It has several implications.[27]

First, it suggests that a theory is 'true' if it works: that is its value. Thus 'the truth' is not an absolute, but changes and grows with time. This reflects war: things that were thought to be true in, say, 1917 are not necessarily held to be true today. That did not matter in 1917! Secondly, that there is therefore no fixed world to be discovered; but rather a quest for workable solutions to our problems. Again, this reflects warfare as we now understand it, particularly given Popper's views. Thirdly, both our knowledge and the world have an evolutionary quality, to meet new situations and needs. This suits us well. The geopolitical situation and military technology change continually. The modern warrior is perpetually faced with new problems.

Pragmatism is intellectually linked to empiricism. Both Isaac Newton and the German theoretical physicist Ernst Mach concerned themselves primarily with the transformed statement (that is, the mathematically general case) 'of *actual facts*':[28] one should observe what happens in real life – what works – and construct limited hypotheses from there.

Pragmatism is relevant to other common disciplines. The workings of the human body are complex; although they are believed to be determined, the full details are not yet known. However, within reasonable limits and the constraints of medical ethics, doctors prescribe treatments on the basis of beneficial outcome. They do not necessarily know the full reason, cause and effect. Similarly, in engineering a full knowledge of underlying theory is not necessarily a barrier to progress. Man built Gothic cathedrals long before Newton developed the laws of mechanics and gravitation sufficiently to enable the structure of those cathedrals to be fully understood.

Pragmatism has obvious application to warfare. A significant aspect of warfare is 'a process of trial and error; seeing what wins, and exploiting it'.[29] Some of modern warfare's more admired exponents showed pragmatism in their methods. Rommel was 'fundamentally a realist'.[30] Both von Manstein and Patton displayed a tendency to observe battle, and then teach their subordinates based on their personal experiences.[31,32]

Empiricism is not just trial and error: it is a logical process based on the structuring of observed facts. Thus an empirical thinker should reject the OODA Loop, since it does not adequately describe the known facts. The empiricist asks 'what happens?', or at least 'what has happened in the past?' Assuming some continuity, he can expect it to happen broadly similarly again. In war he cannot expect causality, but a little continuity taken with pragmatism (that is, concentration on results) suggests a way ahead. We should search history for things that have worked in the past. We can make sensible, if limited, deductions. We can then check them by reference to observed facts. This puts great emphasis on historical study. We should act in war in ways which history tells us ought to have some beneficial outcome: we cannot expect more than that. We should observe the results and then act accordingly. We can find useful insight in theory, but only when it demonstrably accords with the known facts.

Therefore, and critically, military theory should not be a case of 'this is the right course of action', but rather 'doing this will probably have beneficial outcome'. This presents a significant problem. Military historians can almost always present a counter-example – a case where ostensibly similar sets of circumstances (causes) led to different results (effects). Thus in looking at history we should be looking at general trends. Where there are sufficient numbers of examples, we might consider formal statistical treatment of those trends. More broadly, we *must* talk in generalizations. There *will* be exceptions; categoric, determinist rules *will not* be found. Pedants *will* be able to cite exceptions, and thus undermine useful theory. A military historian who rushes to the tired cliché that 'I can show you a counter-example' may be strictly correct but is not necessarily being helpful. One counter-example does not negate statistical significance. And we should remember that historians do not tend to be particularly numerate. Pedants' depredations should be firmly resisted by one simple test: does the theory *generally* aid understanding of useful military problems? If so, then exceptions are permissible. Thus the OODA Loop may be useful in some circumstances and in some applications. It is, however, important to know when those conditions do not apply; and hence what the limitations of the theory are.

A discussion of determinism and its limitations leads to consideration of free will. Determinism is normally discussed in terms of epistemology, the theory of knowledge (hence empiricism, rationalism and pragmatism). Conversely, free will is normally considered an issue of ethics, and therefore the moral rather than the conceptual component of fighting power. Free will is, nonetheless, relevant to warfare in two ways. The first, which we will explore later, concerns the impact of free will on the battlefield. Extreme determinism can be seen as fatalism. If events are predetermined, there is little consequence in human choice. However, determinism is not strong in war. We should therefore take the opposite view. Man can change the future through his actions. Commanders should do precisely that. This is the basis of a philosophy of free will. We should also require our subordinates to express free will. We, and they, should learn to live with the consequences of their decisions. The second aspect is that of the consequence of man's actions. Ethically it is useful (and perhaps essential) to consider man to be responsible for the consequences of his own actions. If he were not, it would be very difficult to live in communities. We should, however, make one observation about the moral aspects of pragmatism. Pragmatism can be taken to imply only that the ends justify the means. Taken to extremes, this can lead to approaches that are morally dubious. However, no reasonable study of ethics would support a standpoint in which the choice of means is entirely justified by the actual outcome. To reiterate: in pragmatism we are discussing epistemology – the theory of knowledge. We are *not* discussing ethics.

If pragmatism is the search for what works, then reducing problems to hard numerical reality is important. For example, it was found during 1915 that the effective neutralization of an enemy trench line required 400lb of high explosive (HE) to be delivered to each yard of trench during the preparatory barrage.[33] Once discovered at Neuve Chapelle, this figure was usefully applied for the remainder of the war. It remained valid even after changes to German defensive tactics (simplistically, a move away from a continuous front line)[34] required the application of that knowledge to be revised.

Similarly there is a reasonably simple figure for the number of bullets required to suppress a man in a trench. It is roughly one round in every 2.5 seconds in an area of about 3 feet square around the area around where his head would be, were he to expose it.[35] Achieving that in practice is not straightforward. A trained British infantry soldier firing single shots at ranges up to 300 yards or so on a range would achieve that every time. In battle there are many indications that this does not happen. Some estimates run to tens or even hundreds of thousands of small-arms rounds fired per enemy killed.

That cannot be entirely true. In north-west Europe in 1944–45, or the Falklands Conflict of 1982, an infantry company might conduct an attack and, after taking some casualties themselves, capture an enemy platoon position. The company would comprise perhaps 90 soldiers carrying 100–200 rounds each. The cost would typically be a few enemy dead; three times that many wounded; and any others surrendered or withdrawn. Even if the company fired every bullet it carried, one enemy would have been incapacitated for each 1,000–1,500 or so small-arms rounds fired. We can deduce that there are some reasonably hard, determinist numbers. However, we need to consider those numbers in their context, and be quite sceptical of their real value.

There is even a harsh, numerical reality in death through old age. In 1999 a newspaper article about veterans of the First World War inadvertently exposed a formula relating to the rate of their demise. Numbers of surviving veterans fell by a factor of 10 for every decade passed (from the age of 60 or so).[36] More routinely, armies used to expect roughly one dead to three or four wounded on the battlefield, given reasonable medical support.[37]

Such numerical observations are useful: the HE requirement served as a sensible rule of thumb for battle planning. They are often found by observation in the midst of war. However, this approach must not be applied *ad absurdum*. War is fundamentally not determined. The ability to apply relatively simple arithmetic ratios to relatively straightforward aspects of battle does not mean that we can usefully seek mathematical explanations of the whole experience. Mathematics describes numerical relationships between entities in an abstract manner. When the fundamental nature of the relationship (that is, the link between cause and effect on the battlefield) is essentially obscure, complex mathematics is at best misleading.

The British have a particularly idiosyncratic notion of 'common sense' which often appears to be synonymous with pragmatism. Common sense is peculiarly Anglo-Saxon: the Germans, for example, have no such concept.[38] Although not generally well defined, common sense refers in some manner to 'commonly held views' of practicality. It was expanded into a formalized philosophical approach by G.E. Moore.[39] Its principal shortcoming stems from the term 'commonly held views'. Views which are commonly held accord with a prevailing world view or paradigm. If the prevailing paradigm is flawed, common-sense views will be misleading. Jomini suggested that most commanders make bad strategic choices because they are misled by common sense.[40] Furthermore, objective truths which conflict with the prevailing paradigm will be seen as 'counter-intuitive'. They will often not be credible, because they do not agree with the common view. Since we know our paradigm of war is flawed, we should not place too much emphasis on 'common-sense' approaches in war.

A significant intellectual criticism of the British generally, and British military thinkers in particular, is that they deride the role of theory. Witness the fact that until 1986 the British Army had no explicit, formalized military doctrine. Such criticism often comes from German and French critics – which reflects their national culture as much as it does the British.[41,42] This may be a fair criticism of British military thought, but it would be unfair of British philosophy generally, and empiricism specifically. Not least, empiricism has its basis in Britain's geographic and strategic situation, and has brought many benefits. Empiricism holds that no hypothesis can ever be proven, merely validated when further reinforcing evidence is found; but it can be disproved. Thus theory is useful, but must for ever remain tentative.[43]

With this as an explicit approach, there is a place for theory in warfare. We can and should develop probable hypotheses about war, and how to wage it. Such hypotheses, or theories, are useful. To Clausewitz, theory was intended 'to guide [the commander's] self-education; not accompany him to the battlefield'.[44] It also calls for the systematic, objective exploration of the lessons of each conflict as (or soon after) it occurs. In general, military theory or doctrine should be a guide rather than a set of rules, since war is not determined to any useful extent.

Note that this is a pragmatic view of theory. William James, the founder of the

American Pragmatist school, considered that theories are effectively tools. In a ruggedly pragmatic sense, hypotheses have value if they work.[45] Theories about war which work – that lead to success in the conduct of battles, engagements, campaigns and wars – are useful and hence valuable. Armies do not get paid to come second, so they want tools which help them come first: every time, if possible.

A further value of theory (or doctrine) lies in codification. Vauban is perhaps the most renowned exponent of siege warfare of all time. Yet it is said that Vauban himself invented or discovered nothing. Every device or technique he used had been employed previously.[46] Vauban did, however, do at least two things that deserve merit. He gathered and wrote down those techniques, so as to make them accessible. Perhaps more importantly, he synthesized a considerable body of disparate method into a coherent whole. Siege warfare after Vauban was far more organized and less haphazard. In practice, seventeenth- and eighteenth-century siege warfare was a highly deterministic process. We might not expect a similar precision in many aspects of warfare today. But we should in some.

Theories involve three kinds of phenomena. They *explain* phenomena which are familiar to us. They *predict* phenomena which we have not yet observed. Yet there is a further class of phenomena which extant theory does not, by definition, explain: anomaly. Anomalies are essentially facts which don't fit the theory. The existence of anomaly causes tension, often manifested in a quest to make extant theory fit the facts.[47] Sometimes anomaly stretches credibility. A revolutionary theory will gain ground if it explains phenomena which extant theory does not.[48] The chief proviso is that such phenomena must relate to meaningful problems. If they do not, the area of thought may simply be ignored – possibly for centuries. Thus in a quest to improve the quality of military thought, we should seek probable hypotheses relating to meaningful military problems. The hypotheses, or theories, which we develop should explain observed phenomena in those areas better than existing theory, if any exists. They should show fewer apparent anomalies. If they do, they will become accepted and commonplace.

Clausewitzian theories of war are a good example. Clausewitz's observations of the Napoleonic Wars did not fit then-current theories or paradigms of war. His theories were largely overlooked, except by Moltke the Elder. Jomini read Clausewitz, but was unconvinced.[49] The outcome of the Austro-Prussian War of 1866 convinced the Prussians. Four years later the French (and to some extent much of the rest of the World) were forced to accept a Clausewitzian view of war by the events of the Franco-Prussian War. After 1871 Clausewitz's thoughts were widely considered to be conventional military theory.

Two other classes of theory are of interest. Conjecture or hypothesis which is found to be wrong is often useful. The process of exploring and attempting to validate it leads thinkers to explore their problems in a structured manner.

Such rejected theory often arises when a discipline is in a state of flux.[50] In these conditions extant theory has distinct shortcomings, and practitioners are casting about for better theories. Popper actively chose a method which prompted continued discovery, even though some of the theories which might result could be wrong.[51] Understandably, soldiers tend to be more cautious.

The other class of 'wrong' theory is theory which has been superseded. It is often useful, as long as its limitations are recognized. Newtonian mechanics is an example. Einstein's discovery of relativity showed that Newtonian mechanics is actually a highly limited special case. It is an approximation which only applies with any accuracy in very limited circumstances: specifically, where the velocity of an object is very much smaller than the speed of light. However, those circumstances describe all meaningful motion on the surface of the Earth.[52] They apply to all building projects: buildings are not intended to move at all, let alone at an appreciable fraction of the speed of light! Thus Newtonian mechanics may be 'wrong', but pragmatically it remains useful.

The OODA Loop is a further example. It is a poor model; but for several years it gave soldiers a descriptive framework on which to hang discussions of command and control. It did not adequately describe the process of defeating an enemy, but it still describes a simple process of decision-making. To that extent it remains useful, as long as its limitations are recognized.

The technique of evidence-based practice is a further refinement to limited empiricism and pragmatism. Forty years ago the practice of medicine in Britain was a combination of patriarchal teaching methods and received best practice. Doctors performed techniques largely because they had learned to do so; often at the hands of doctors who themselves did so because *they* had learned to do so. Evidence-based practice was introduced in the 1990s to ensure the relevance of technique as actually practised. It is also part of a continuing quest for development. The doctor is required to undertake procedures only if he is aware of the evidence of their efficacy. This makes him examine the evidence and test its validity. He may find that the evidence is scant, nonexistent, or overtaken by developments elsewhere. If so, he is obliged to reject the technique for something better.[53] Across the body of medicine, evidence-based practice drives a continuing quest to refine, update and improve.

Medicine may be particularly amenable to evidence-based practice, not least because the sheer number of patients and individual treatments involved allows relatively easy statistical treatment of observations. Warfare may be more problematic. Even in a major war, a soldier may have difficulty in assembling a statistically significant body of experience. He will normally have other things to attend to (such as staying alive) than conduct statistical analysis. In peacetime we might reasonably expect him to examine history, and experience gained in training, and then actively question whether the evidence available to him supports his current practice. Where possible, he should search history for broad, statistically significant trends.

Furthermore, the habit of enquiry would help overcome what T. E. Lawrence described as 'the fundamental, crippling incuriousness of the British Officer'.[54] Lawrence's criticism is somewhat harsh. However, any army would be better if its officers displayed a more questioning attitude to their profession. Army officers may be many things, and many of them are admirable. But, in my experience, they do not tend to be professionally inquisitive.

The British traditionally regarded warfare as being divided between the strategic level (the waging of wars) and the tactical (the fighting of battles). In the late 1980s British doctrine recognized the operational level of war, being the conduct of campaigns. By 1995 or so it described the grand or national strategic level as the pursuit of the objectives of the state at a national level. The military strategic level consists of the formulation of military objectives and the provision of resources by which to achieve them. Those objectives are normally expressed in terms of campaign directives, by which one or a series of campaigns are initiated with the desired end state of achieving national goals. The operational level of war is seen as the sequencing of battles and engagements, and the concomitant provision of resources, in order to achieve campaign end states required to meet strategic objectives.

The tactical level of war is therefore ostensibly straightforward. Forces and other resources are committed to battle at specified times and places as part of a campaign plan. Yet this conceals two significant aspects. First, it may be ostensibly straightforward, but it is without doubt difficult: Clausewitz considered that '[e]verything is very simple in war, but the simplest thing is difficult'.[55] Secondly, fighting is the currency of war.[56] Troops and resources may be 'committed to battle', but what they do is fight. Battles and engagements do not just occur. They are fought. Campaigns can only run to plan through victories, and victories are bought through fighting. '[S]trategic opportunity, as the saying runs, follows tactical success'.[57] More generally, fighting is not just the definition or common denominator of war; it determines its whole character.[58]

This raises an important issue: victory! As a further example of the poor state of military thought, there appears to be no consensus as to what constitutes victory. For example, it might be commonly held that the Western Allies were the victorious side in the Second World War. Conversely, a historian could point out that in 1945 Poland was occupied by a foreign power, and that Poland's freedom had been a war aim in 1939. He might also point out that by 1955 or so the German economy was generally in better shape than the French or British. Surely Germany had won, to that extent?

Such observations are valid, and the absence of consensus as to what constitutes victory points once again to a poorly developed paradigm. The problem may arise from failing to differentiate between success at the tactical, operational and strategic levels. The last two are generally dependent on diplomatic and even perceptual issues, whereas the first is generally more clear-cut. Tactical success – success in a battle or engagement – is usually easy to identify, especially in relation to the protagonists' goals. Those may well change during a battle. But 'success in battle' is generally easier to understand than the wider aspects of victory.

Battles are fought, and campaigns are waged through their results. Campaigns are waged, and thereby strategy is conducted. Operational art and strategy are primarily concerned with resource allocation and event sequencing. Decisions concerning both aspects must be made, normally in conjunction with each other. However, in both decisions are not normally made in 'real time'. Equally, the consequences of decisions are not normally visited suddenly, dramatically and possibly fatally on the decision-maker.

However, that does occur in battle. At the tactical level, decisions are often made in real time. The consequences may be lethal for the decision-maker and his immediate neighbours. War 'requires people to act in the most literal peril of their lives'.[59] Although military decision-making at the operational and strategic levels is sometimes pressured and often of weighty consequence, it does not have the dynamism and suddenness of consequence of tactical decision-making. Since fighting is the currency of war, tactical decision-making is a key, and perhaps *the* key, activity in warfare. It is the mental activity which most directly affects the outcome of combat.

This applies equally to irregular warfare. In such operations, campaigns can be envisaged as a linked series of confrontations with the other protagonist(s). Those confrontations or 'engagements' usually include the manipulation of the threat of armed force. Tactical decision-making in such operations may be just as real-time, and perhaps just as lethal. The key difference is that the probability of violent outcome is generally somewhat less.

Until perhaps the 1970s there was an unspoken assumption that somehow battles just happened: 'Napoleon did this and therefore ...' This is partly a consequence of a style of military history which focused on the personality of the great commander, and glossed over the mechanics of the battlefield. However, subordinate commanders make decisions and give orders; units manoeuvre; the respective sides clash; and an outcome emerges. Of course, the process is far more iterative and complex than that. However, observation tells us that battles are actually *fought*. They do not just occur.

It was perhaps not until the French campaign in Egypt in 1799 that an army quite explicitly considered which tactical methods to use in a given campaign, rather than simply 'when' and 'where' to fight.[60] Thus over the last 200 years or so, man has (more or less explicitly) designed weapons, developed tactics and made battlefield decisions. Of those three activities, the one that must be conducted on the battlefield is tactical decision-making. Designing weapons and developing tactics are best done out of contact with the enemy. Tactical decision-making cannot be.

Tactical decision-making is unique. Only in war is decision-making routinely of lethal consequence to many, whatever the outcome. Only at the tactical level must decision-making be carried out in real time. Thus it differs from non-military decision-making, which is not routinely lethal in its consequences. It differs from operational and strategic decision-making, which may well be lethal in its consequences but which is not so time-pressured. That is not to denigrate the importance of the operational or strategic levels of war, but simply to assert that tactical decision-making is a key mental process. It defines the scope of this book, which considers the environment where that decision-making takes place, its impact and the skills required. Since decisions must be taken in real time, and on a battlefield which is dangerous, battlefield decision-making will be conducted under stress. Its output will typically be sub-optimal if viewed objectively.

There is perhaps a narrow definition of the battlefield as a place of high-intensity conflict, which occurs rarely and only in major wars. Were that so, battlefield decision-making would be of limited utility. Such conditions are, however, very much in the eye of the beholder. They are more common than may at first sight seem apparent. Recent operations in Bosnia, Rwanda, Somalia, Iraq or Afghanistan might have been characterized as peace-support operations. To an infantry section commander coming under fire in any one of them, his experience was in all important regards that of war. Here we do not differentiate narrowly between regular and irregular warfare. Instead we focus on any case where decisions must be made in real time under conditions of lethal stress.

Warfighting is fundamentally a human activity, in which humans choose what to do, consciously or subconsciously; rationally, irrationally or non-rationally. Fundamentally, three things occur on the battlefield: men think, move and commit violence. All other activities support those functions. There is a huge premium in applying violence at the right time and place. Deciding where and when to fight is very important. Therefore tactical decision-making is a key activity.

It is the key *mental* activity. It is not necessarily 'the key activity'. One could argue that for the Israelis in the Six-Day War their key activity was tank gunnery. One could argue that technical and numerical improvements were the key activities in turning the tide in the Battle of the Atlantic in 1943. Equally one could say that Rommel delivering supplies to his panzers was the key activity in the fighting in the Cauldron in the Gazala battles of mid 1942. However, these are fundamentally not *mental* activities.

3

The Nature of Combat

Current intellectual methods are generally inappropriate. We should, instead, use an approach based on limited empiricism and pragmatism. But what is war, combat, fighting actually like? What is the phenomenon we are dealing with? We should start by considering the shortcomings of the current view of combat. We face a significant difficulty. Our view or paradigm is not explicit and shows little coherence. To explore it in any simple manner is to prompt objection: it is unlikely that any two readers will have the same view of combat, if they have any explicit view at all. Trying to describe the paradigm directly is likely to fail. Nonetheless, we can make some broad initial statements. We have already observed that our current description of combat contains several poorly described phenomena. What is shock, surprise, or 'the initiative'? Some extant theory, such as the OODA Loop, is an inadequate description of the behaviour it addresses.

The current paradigm is infested with paradox such as 'defence is the best form of attack'. We can also find examples of counterintuitive behaviour. For example: despite 50 years of advances in automotive design of AFVs, the rate of advance in the Gulf War in 1991 was no faster than the fall of France in 1940.[1] The German 2nd Panzer Division advanced from Sedan to the English Channel in May 1940 at about 45 km per day. In March 2003 the 3rd US Infantry Division (mechanized) advanced to Baghdad at about 115 km per day from the Kuwaiti border to An Najaf. However, assuming perhaps 12 hours' movement per day, that corresponds to only about 10 km per hour. And if we consider the Division's advance to Baghdad overall, the average speed was only 50 km per day or so: eerily similar to France in 1940. Automotive speed is also, clearly, not a major factor. We can also find behaviour that our paradigm doesn't even acknowledge, let alone describe. For example, some fighter-pilots appear to have been far more effective than the average. So do some submarine captains, tank commanders and snipers. Is this a discrete phenomenon? If so, should it be exploited?

We can easily find other weaknesses in our accepted view of combat. For example, there is little if anything in British military practice which suggests that combat is adversarial. Tactical instruction tends to concentrate on the mechanics of low-level skills and drills. Training pamphlets reinforce that bias, with very little to suggest what the impact of enemy action on a procedure undertaken in war might be.[2]

Collective field training is usually constrained, with a controlled enemy, so that the 'right' lessons are learnt from the exercise. In part this is sensible: in the early stages of collective training it is important that a manoeuvre is completed so that the troops can see what should happen. Unfortunately that makes it a very unrealistic simulation of what happens in combat which *is* adversarial. Command and staff training as practised at purpose-built training establishments is explicitly aimed at practising staff procedures, not on beating the enemy. Again, this may be sensible at a certain stage, but it falls short of training to beat a live enemy. So why not train to win? Is process more important than winning?

Similarly, and perhaps surprisingly, the British Army appears to have forgotten about speed and tempo. This is an interesting case of the difference between espoused and enacted behaviour, and makes a useful case-study. Achieving *superior* tempo relative to the enemy was endorsed in doctrine in the 1990s, and this appears sensible. However, the same doctrine suggested that one should be able to *impose* a tempo on the enemy, and that that superior tempo is not necessarily faster.[3] This is clearly rubbish. If combat is adversarial, then 'the other guy gets a go, too'. Imagine a boxing-match in which one boxer attempts to move and punch more slowly than his opponent! If you go slowly, you simply give your opponent time to react. His reaction might quite sensibly be to pre-empt you – that is, go faster. In an adversarial contest, superior tempo is about going faster.

The one exception is that of slowing the enemy down, in order that you can generate or sustain an advantage in speed. Destruction of the enemy's C2 networks, attrition of and delay to his forces may all slow the rate at which he can decide and act. However, the aim is still to ensure that you can act faster. Thus although superior tempo is *relative* to the enemy, it remains *relatively faster*. Furthermore, one may wish the enemy to decide and act more slowly, but one cannot guarantee what effect that will have.

But how fast is fast? 'As fast as you can' is the only safe response. The discussion is nugatory until one makes comparisons. In the Second World War, divisions normally made plans for their brigades or regiments to carry out the next day. This typically means planning for a period which starts about 6–12 hours ahead, with some idea that the envisaged action would last about 12 hours (since few battles were actually conducted around the clock). Today, with the 24-hour battle an apparent reality since the Gulf War, doctrine appears to require divisions to plan not 6–18 hours ahead but 24–48 hours.[4] Thus the decision–action cycle is actually longer, and the rate of tempo which a division can achieve thereby *reduced*.

There are three further general criticisms of what appears to be the British paradigm of combat. The first is a reliance on mechanical analogy. Fuller wrote of the 'shot in the head', and Liddell Hart of the 'expanding torrent'. Both are entirely respectable analogies and suitably graphic. However, what is missing is any suggestion of philosophical generalization. Is there any concept to compare with Clausewitz's dialectic of aims and means?

Secondly, we tend to judge by what we can observe most directly, and what we can most easily measure. For example, Command and Staff Trainers practise staff procedures, not least in part because procedures can be checked using a tick-list. It is far more difficult to judge whether a plan would have succeeded against a real enemy. 'Winning' against the computer-controlled enemy, or losing, is almost disregarded.

The final criticism is the absence of any concrete link between military theory and human behaviour. Military men stress the importance of the human dimension in war, and seem right to do so. Much of British military practice revolves around intensely human aspects such as the regimental system. Yet there is almost no intellectual linkage between human behaviour and military activity. One looks largely in vain for any reference to theories of motivation, human interaction or even decision-making in doctrinal publications. It is almost as if such issues do not exist.

Thomas Kuhn pointed out that novel paradigms are often obviously flawed to those who adhere to the preceding paradigm. For example, to a pre-Copernican astronomer it was patently obvious that the world did not move: that wasn't the sort of thing that the world did.[5] Similarly, to a Newtonian physicist it was patently obvious that space isn't curved – space just wasn't that sort of a thing. However, scientists now generally accept that Einsteinian space is curved. This is far less obvious to the average inhabitant of the world today, so Newtonian dynamics persists as a paradigm. Two aspects of the idea of paradigms deserve comment.

First, what is the value of a paradigm? A paradigm is the way in which we view the world. A better paradigm should allow a better understanding of the world around us, and explain a greater range of relevant phenomena. That should allow us to develop more useful theories. Useful theories help us solve meaningful problems, and perhaps predict behaviour. A better paradigm also allows us to resolve paradoxes.

Secondly, it is quite understandable for a military writer to say that doctrine isn't the sort of thing that considers human behaviour. That's the realm of the moral component: the stuff of leadership, morale and so on. But that is flawed. We seek to understand combat, which involves thousands of human beings reacting to very stressful circumstances. They are not automata. We cannot attempt to understand combat in any meaningful way without understanding human behaviour.

This may be a lot to ask, particularly in an environment where determinism is not strong. The theory of hydrodynamic flow is a useful analogy. It describes the mechanism of the penetration of armour by solid projectiles at very high speeds (typically above 1,000 metres per second). The theory suggests that the projectile effectively flows through the armour as if it were a fluid. This defies our understanding of the properties of matter: that isn't the sort of thing that solids do. Critically, no one suggests that it actually happens. Yet the mathematics that results from *assuming* it does accurately predicts the performance of armour-defeating projectiles in many circumstances. The theory of hydrodynamic flow is therefore useful. So, even if we are not entirely comfortable with their content, theories set within appropriate paradigms can be useful.

Our current paradigm contains an abundance of paradox, a range of poorly described phenomena and incoherent theory. It lacks adversarial content, and largely overlooks tempo and human behaviour. Yet we fight our wars that way. Given that we also tend to start wars the way we finished the previous one, that is like driving a car blindfold, guided by a passenger looking out of the back window. Alternatively, it is like playing a team-game where the rules have changed, and nobody can remember what the rules were. We can, and we should, do better. We should develop a better paradigm which more accurately describes combat, resolves paradox and helps us develop useful theories of how to fight.

In sound empirical fashion, we begin with what we can observe. At first sight combat can appear 'formless, unstructured, unutterably confusing and often dangerous'.[6] If, however, we step back from the participant's perception we see that combat involves fighting, but is more than a series of one-on-one fights. Combat has higher-level characteristics, and is conducted in accordance with higher-level aims. The soldier rarely fights because he wants to, although he may. Typically he fights in a given place and time as a consequence of higher-level orders. The aim is rarely to kill or incapacitate individual soldiers, although that usually occurs. In some manner combat is an assault on the enemy force as a whole. Similarly, it is not settled by the resolution of a series of individual fights. At some point, one side (or occasionally both) desists and withdraws from combat, sometimes with dramatic effect.

We appear to be observing complex and perhaps chaotic behaviour. Clausewitz twice repeated Napoleon's dictum that the complexity of the problems involved would require a Newton to solve them.[7] Research into complexity and chaos has occurred in the last 20 years or so, aided by developments in computers. The fundamentals of complexity theory consider the behaviour of large numbers of elements and the relationships between them. The way that those elements are organized results in lots of feedback loops. Any system with a large number of elements and feedback loops displays emergent properties or behaviour. Such 'emergent behaviour' is not in practice predictable in advance from the characteristics of the individual elements. Nor is it predictable from an understanding of the relationships between any two elements.

This appears to apply to combat. At any level (for example, that of individual combat between soldiers; or between armoured divisions) we can say much about the individual elements (the soldiers, or the divisions). We can say a certain amount about the interaction between those elements. For example, the relationship between an Israeli and an Egyptian soldier (which is better trained, which is better motivated, hence which is likely to come off better); or between an Israeli armoured division and an Egyptian one. But we have great difficulty in saying much about the outcome of a battle as a whole. Indeed, we can say that only past experience of very similar circumstances is any guide at all, and that can be misleading.

Complexity theory is so narrowly applicable that what scientists describe strictly as chaotic behaviour has never been observed in the real world.[8] Nevertheless, since it gives enormously powerful insight into observed behaviour, complexity theory is used in several important applications. Weather forecasting is one. The American meteorologist Garnet Williams provides a very useful description of many aspects of complex systems:[9]

- They comprise a large number of somewhat similar but independent items, particles, members, components or agents.
- The system contains dynamism: each element continually acts or adjusts to its fellow agents in perpetually novel ways.
- The system adjusts to novel situations so as to ensure survival, or bring about some advantageous alignment.
- The system is capable of self-organization, whereby some order forms inevitably and spontaneously.
- There are local behavioural rules that govern each element.
- There is a hierarchical progression in the evolution of rules and structures. As evolution goes on, the rules become more efficient and sophisticated, and the structure becomes more complex and larger.

This appears to apply to several aspects of combat. We can usefully develop a limited numerical understanding of complexity. Consider a group of 'n' elements. Let any or all of them exist in up to a total of 'm' states. The total number of possible states in which the overall system could exist, 'p', if any element can be in any state, is given by the expression

$$p = n^m$$

This appears innocuous. However, the power function (n *raised to the power of* m) has huge consequence. For example, take a platoon of 30 soldiers, the individuals of which could at any one time be on leave, at work, off duty, on guard or in gaol. That is five states, so 'n' = 30 and 'm' = 5. Hence 'p', the total possible number of states of the platoon, is 30^5 or 24,300,000. The number of links between members of a group of people (and hence the number of relationships between them) goes up roughly with the square of the number of people involved. The number of relationships grows fairly rapidly, but not as fast as in the power function above. For our platoon, the number of relationships is about 450.

Things become far more complex when we consider the number of possible states of all those relationships. If each relationship can be in several states (perhaps friendly, senior to, paid more than, or whatever) then those 450 are themselves raised to a power. For just ten possible types of relationships, the possible complexity of our platoon is 2.426×10^{26} – an inconceivably large number. In the real world the relationships between 30 people who work together are always changing. Which one of those 2.426×10^{26} possible combinations represents reality at any moment? (Little wonder that platoon commanders often feel harassed!)

The purpose of all this is not to scare the reader with large numbers. It is to suggest that the number of possible 'states of the world' for even a small part of an army is vast. If there is any change in the system at all, the number of subsequent states is also vast. Such arithmetic is described as a 'combinatorial explosion', because it produces a very large number from a combination of a relatively small number of elements.

This begins to suggest why combat may not appear to be determined. The number of possible states of a system so complex as an army is inconceivably vast. The number of states for two armies locked in combat (with thousands of soldiers fighting thousands of their enemies, in ways that can vary considerably) is yet bigger. The actual and perceived complexity of a battle will be utterly inconceivable, and our ability to predict the precise outcome will be effectively nil. There will in reality be only one outcome to a battle, but the probability of predicting with precision what that outcome will be is one, divided by the number of possible outcomes: an inconceivably slim chance. No punter would bet on odds of 2.426×10^{26} to one.

Complexity theory does not attempt such predictions. Instead, it makes useful observations about the overall behaviour of such a system. For example:

- The more complex the system, the further away cause and effect usually lie from each other, in both space and time.
- It doesn't take many feedback loops before it becomes hard to predict the behaviour of the system.
- In any given system there are very few high-leverage points where one can make lasting changes in the overall behaviour of the system.
- Neither the high-leverage points nor the correct way to move those levers for the desired results tend to be obvious.
- 'Worse before better' is often the result of a change at a high-leverage point in the 'right' direction.[10]

A very insightful deduction is that, in complex systems, constant experimentation and 'try it and see' is highly preferable to 'tell me what I can do to fix it'.[11] An empirical rather than deterministic approach is preferable. We saw earlier that war may be determined, but it is not useful to consider it to be so. Cause does not appear to lead strongly to effect in combat, and we now begin to see why.

Complexity theory describes systems of numerous but semi-independent elements which interact. Such systems can generally exist in one of three states. Where the elements do not interact very much, the system is in equilibrium. The behaviour of the system is moderately predictable. As the rate of interaction increases, the system changes into a 'complex' state, in which patterns and emergent properties can be observed. In this condition the future state of the system cannot be predicted with much accuracy. Finally, as the level of interaction increases yet further, the system becomes chaotic. The behaviour of the system is so turbulent that human understanding and intervention become impossible. Only gross measures (such as withdrawing all elements from the system, for example by withdrawing from battle or surrendering) have any obvious effect.

Thus modern combat appears complex, and at times verges on the chaotic. Interestingly, the Israeli historians Luttwak and Horowitz described the Israeli army as 'suspended between dynamism and chaos'. This was written in 1975, before the advent of Chaos Theory.[12] We might conjecture that it was the abandonment of close-order drill on the battlefield in the nineteenth century that endowed combat with its complex and chaotic character. At a higher level complex behaviour was in fact observed in the Napoleonic wars – particularly in the actions of the French army in the Revolutionary period.[13]

The ability of an element to survive in a complex and hostile environment is affected by its ability to adapt, by making the right decision at the right time. There is a vast number of possible future states of a complex system. Some of these will be known; others are in principle unknowable. That is because the amount of information which an element would need to know in order to ensure that no potential outcome is a surprise is vast. Some outcomes are unknowable, and some unknowable outcomes are undesirable. They are called 'rogue outcomes'. An element, entity or organization can normally learn a set of useful outcomes with time; but there will always be some rogue outcomes. Organizations can usually accommodate this by adapting structure and process for greater precision, particularly at the part of the system where the organization is located. But that is not the only strategy: multiple redundancy is another. We will look at such strategies later.

Before leaving complexity theory, we should look at two further implications of combinatorial explosions. Firstly, the number of relationships rises roughly with the square of the number of elements in the system. Reducing the number of elements slightly does not do much to reduce that aspect of complexity. In order to reduce that aspect significantly, one has to go to much smaller numbers. For example, if we consider a platoon as three sections, rather than 30 soldiers, the complexity at platoon level is much less. Good platoon commanders learn that quite quickly. Organizational subdivision can assist the management of complexity. Secondly, some simple examples show that reducing the number of states has far more impact than a comparable reduction in the number of elements. That is because interactions grow with a square law, while total states grow in a power law.[14] Thus to reduce the real and apparent complexity of an organization, divide it into a small number of units and manage it in ways which can be described by relatively few states.

This indicates how complexity is managed in practice. Consider a division with, say, 12 battalion battlegroups in three brigades. We can describe a battlegroup as the basic building-block from the divisional HQ's perspective. If the potential activities of each battlegroup are listed in doctrine as (say) attacking, defending, delaying, advancing or in reserve (a total of six states) then the division can exist in 12^5 or 258,832 possible states. That is a large number, but very much smaller than the 2.426×10^{26} states previously considered for the platoon. Furthermore, as we shall see, the division rarely commits more than four battlegroups in its first echelon. The remainder will typically all be doing the same thing. For example, if the division is withdrawing, the first echelon will typically delay the enemy, whilst the remainder retire. This reduces the potential variability of the division enormously – to something like $4^5 \times 5$, or 5,120.

Considered this way, the division is much easier to comprehend than the platoon. There are two reasons. Firstly, the number of elements is smaller (12 versus 30). Secondly, no attempt was made to capture the actual relationships between the battlegroups. Tools such as subdivision (of a division of many thousands of men into a few battlegroups and brigades) and global activity descriptions (the tactical operations: advance, attack, delay, etc.) allow astonishingly complex organizations to be managed effectively in practice.

Thus far we do not have a paradigm of combat. We have an observation that combat is like a complex system that borders on the chaotic, and some understanding of such systems. We need a model of one element, or a group of elements, in order to give more insight. Consider a soldier, for example an infantryman. He can basically do two things which affect the enemy. He can move across the battlefield, or he can fight the enemy (if we include actions such as firing, throwing grenades or bayoneting the enemy within 'fighting'). He also has to decide where and when to move, and who to fight.

Now consider a unit, say an infantry company. In terms of its effect on an enemy, the company commander can also do two things. He can move the company or have it fight. He also needs to decide where and when to move it, and what to fight. Unlike the infantryman, it is not himself that he moves (although he does) so much as his platoons. Similarly platoon commanders move their sections, and battalion commanders move their companies. Hence we begin to see a nested series of elements. In all of them the commander (perhaps with a headquarters) makes decisions which result in a group of subordinates moving, fighting, or both. Thus a company might be described by Figure 1.

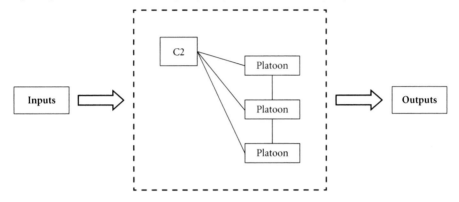

Figure 1: Basic System Model of a Company

The company has inputs, principally in terms of superior orders (to the company commander) and logistics (which he directs to the platoons and to a limited extent to his HQ). It has outputs which (in terms of combat) are ammunition expended, casualties inflicted and sustained, and ground or enemy positions captured.

The principal interaction between the C2 element (the commander and his HQ) and the platoons is information, in the form of orders downwards and reports upwards. The interaction between the platoons may include information but also includes movement relative to each other, covering fire, and so on. The company model is more complex than shown, because the company commander can also ask for fire support from elsewhere. He may have other elements grouped within his company, such as machine guns or anti-tank weapons. Figure 1 is, nonetheless, a useful first approximation. Now let us look at a battalion, as in Figure 2.

Figure 2 can be seen (at this level of abstraction) to be the same as Figure 1, with the same inputs and outputs. Company models could be shown nested within the battalion model. Similarly a platoon can be seen as a C2 node with a nested set of sections, and a brigade as a C2 node with a nested set of battalions.

The behaviour of any element within any of these models can be described, within reason. We can read in tactical handbooks what a company, platoon,

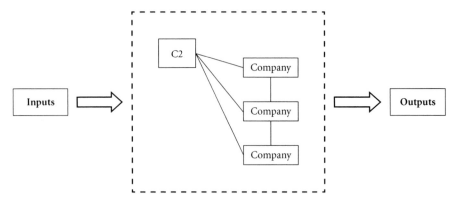

Figure 2: Basic System Model of a Battalion

battalion or brigade is supposed to do. We can say with some confidence what happens when (say) a company attacks an enemy platoon, or defends against an enemy battalion. Thus, in complexity terms, the behaviour of an element at any level of decomposition is understandable within reason.

This set of models represents a nested set of systems, which is well described in complexity theory. Fractal patterns are fundamentally what occurs when certain types of behaviour recur over several levels. For example, the sinuous nature of a coastline can be observed when the geography is decomposed from the level of a continent to that of a single grain of sand. Such behaviour is not, however, perfectly isotropic: there are variations as you go up and down the scale. A corps does not behave like a platoon. The behaviour of the divisions of a corps can be described in broadly the same way as the sections of a platoon: move and fight. There are, however, differences of scale, and other differences which are typically scale-dependent.

We are on the edge of General Systems Theory, first described by the Austrian biologist Ludwig von Bertallanffy.[15] We will consider Systems Theory later. Here we will say only that although we can draw Figures 1 and 2 as systems with fairly precise boundaries (the outer dotted lines), few human systems are so clearly defined in reality. There are blurred boundaries: which 'system' does an attached artillery Forward Observation Officer belong to? Hence rarely do real systems have categoric, black-and-white, sharply defined behaviour. It is only when a system is considered in its environment that a useful picture of its behaviour emerges.

We do not have to believe that complexity theory is 'true'; only that it gives useful insight into combat. Certainly at this fairly modest level, complexity theory is not likely to provide 'the answer'. Hopefully it provides useful insight into factors and processes.

Thus far our description of combat is basically a series of nested elements interacting with the environment, with each other and with the enemy. Uncertainty will be inherent, due to variability between elements, relationships, scales and the emergent behaviour of the system as a whole. Complexity theory suggests that, at best, we can be weakly predictive. We can suggest that certain kinds of behaviour will have the right sort of effect; but we cannot say with much certainty what the overall effect of that behaviour will be. War may be determined, but it is not useful to consider it to be so.

The process of modelling combat is relatively common in defence research establishments. Jonathan Searle of the Royal Military College of Science at Shrivenham produced a description of combat as being: the realm of human behaviour and decision-making; two- or more sided and competitive; highly complex; unstructured, hugely concurrent, highly dynamic and rapidly evolving; having a range of important time bases, and very difficult to validate and verify.[16] This now seems familiar. We have a model of combat in which:

- Overall, the numbers of elements involved, and the interactions between them, make any attempt at detailed prediction meaningless.
- Each side consists of a nested set of systems. Each has a C2 node and a number of subordinates. The C2 node makes decisions and, simplistically, orders subordinates to move and to fight.
- At each level, information, orders and logistics are inputs. Outputs include casualties inflicted and sustained; ground gained and lost, and enemy positions captured. Additionally, firepower can be injected from elsewhere.
- Combat is adversarial: the outputs from one element become inputs to an enemy element. Casualties inflicted by one side are sustained by the other. Positions captured by one side are lost by the other, and so on. This process is a many-to-many relationship. A defending platoon may inflict casualties on a number of attacking platoons, and an attacking company may capture positions belonging to a number of enemy platoons in one attack. Thus the interrelationships between opposing forces are intensely complex. They will seem to be unutterably confusing to an observer on the battlefield.
- The behaviour of each element at each level is moderately predictable. However, the overall performance of two forces in a battle or engagement is not. Nonetheless, the fact that at each level the behaviour is at least to some extent predictable does allow some control.

We continue in an empirical vein: what can we observe? We are trying to develop a necessary and sufficient description of combat.[17] To do so we must look rather more closely than the popular view of war. Our perception tends to the superficial. The layman might suggest that since the 1940s war has been dominated by the tank and the aircraft. Conversely, a detailed study of war from 1973–89 identified 23 major lessons, among which the more important were:

- No force on any side of the five conflicts under study[18] was able to meet its own expectations regarding its technical and operational ability to integrate combat arms.
- No single area of tactics, technology and readiness ultimately had more critical overall impact than infantry combat.
- In each case, however, CAS [Close Air Support] activity had far less impact on the ground battle than either the participants or outside observers anticipated.
- In general, long-range bombing rarely had the anticipated effectiveness.[19]

In short, combined arms tactics were not as good as people thought, CAS and air interdiction were less effective than assumed, and in the long-run much depended on the quality of the infantry. War is not as it appears.

Another example comes from the Battle of Kursk. A detailed numerical analysis suggests that our perception of the losses incurred by the Germans is considerably inflated, mostly by Soviet propaganda. Relative to the force ratios involved, the Germans performed quite creditably. The battle was certainly not the 'death ride of the German panzer divisions', as it is often portrayed. Indeed, losses of Russian tanks were far more significant than were German losses.[20] When we observe the 'facts' in war, we must be careful that we are actually seeing the war 'as it actually was', as Leopold von Ranke, a contemporary of Clausewitz's, put it.[21]

We have stressed that combat is a human activity, and we begin to see how. The human is an actor in the process. Individuals make choices in the systems described above. The commander can choose how, where and when to participate. At the lowest level, the individual element is the soldier. In the extreme, he can make any of those choices for himself. He can refuse to participate. He may be astonishingly brave. Or he may, like the great majority of soldiers, take a middle path conditioned by things such as his sense of duty.

The human, as an agent on the battlefield, can make rational choices. It is rational to avoid the enemy's strengths, particularly his lethal ones, and attack his weaknesses; particularly those that reduce his ability to commit violence. We should expect the same in response. Conversely, soldiers may not always act rationally. Surprise, shock and fear all affect their behaviour. The possibility of one or a number of soldiers acting irrationally is always present. It is a likely consequence of a dangerous environment. This considerably increases the number of states which any one battlefield element may display, which again hugely increases the number of possible outcomes of an engagement. It is another reason why combat is most unlikely to be predictable in advance.

Anecdotally we know that human issues are hugely important. Fear, courage, bravery, cowardice, leadership, discipline, ethos and *esprit de corps* figure highly in all but the most superficial accounts of battle. Military doctrine tends to acknowledge them, but does not seem to link such observable human behaviour with tactical outcome. However, the human is not just the victim of his emotions. He is an agent on the battlefield and he influences the outcome of battle by his action, to a greater or lesser degree. His viewpoint or understanding of battle is highly important. Since combat is so complex, the participant needs to explore his environment. He should use force – attack – if necessary to enable him to find out. After all, the enemy will want to stop him doing so. This has several aspects.

Firstly, 'finding out' enables him to make better decisions, based on better and more recent information. The converse, waiting for better information to arrive, is doomed in an environment as dynamic and adversarial as combat. Secondly, attacking makes things happen. It influences the situation in a way that is generally to the attacker's advantage, or at least in ways which he can understand (since he initiated the activity). A classic example is the action of General Hans von Gronau, commanding IV Reserve Corps on the march to the Marne in August 1914. His was the right rear corps of von Kluck's First Army, and hence the right flank of the entire German army in the West.[22] His right flank had been harassed all morning by French cavalry. In the absence of information as to French dispositions or intentions he attacked – and caught three French divisions deploying for an attack themselves. By doing so he probably saved the German First Army.[23]

By going and finding out, and making things happen, one can influence the battlefield. Combat is uncertain, and will tend to reward those who can tolerate uncertainty. Tolerance of uncertainty is the hallmark of pragmatism. General David Fraser wrote of Rommel's experience that 'war is so uncertain business, so dependent on a concatenation of unpredictable chances, that boldness, a touch of optimism and above all speed can and generally should do better than attempts at exact calculation'.[24]

This reinforces the importance of humans as agents in combat. 'Warfare is more than [weapon] systems; it is fundamentally and ineluctably a contest of human wills.'[25] The business of going and finding out, of making things happen and influencing the battlefield can be quantified. In a study of 158 land campaigns from 1914 onwards, David Rowland and his colleagues at the British Defence Operational Analysis Centre found that just three or four battlefield factors dominated the probability of success at the campaign level. One was surprise. The second was possession of air superiority. Air superiority enabled freedom of manoeuvre for one's own ground forces and permitted some interference in the enemy's movements. The third was 'aggressive ground reconnaissance'. This meant aggressively exploring the battlefield, and exploiting weak spots and opportunities as they were discovered. The fourth was shock. All four factors were individually equivalent to large multipliers in force ratio, in terms of impact on campaign outcome. They were far more important than factors such as force density and achieved force ratio.[26]

The achievement of initial and multiple surprise has the same effect as a force ratio of 2,000:1 on average, in terms of their impact on achieving a breakthrough at the beginning of a campaign. 2000:1! It also has the same effect as a force ratio of 260:1 (on average) in terms of overall campaign impact. These figures are averages. But in 95 per cent of all occasions where surprise was created, the effect was *at least* as great as that of a force ratio of 10:1.[27]

The aggressive exploitation of reconnaissance results in:
- the seizure of opportunities
- the neutralization of enemy reconnaissance
- the location of gaps in the defence
- disruption to the defence, including in rear areas and HQs
- demoralization
- physically threatening rear areas and HQs
- creating uncertainty, through destruction or capture
- disguising the nature and direction of the attacker's thrusts.[28]

None of these relate to the attritional reduction of the enemy's force. They all relate to the enemy's will and cohesion. Overall, aggressive reconnaissance has the same effect as a force ratio of 26:1 on average, in terms of its effect on campaign outcome. This is highly significant. Force ratios of 26:1 are almost impossible to achieve. Conversely, the use of aggressive ground reconnaissance has occurred quite often.

These statistics are highly important for two reasons. Firstly, they are highly important in terms of approach. Given that typical force ratios were typically between 1:1 and 6:1, they show that what leads to success *is not numbers, but things that upset the enemy's will and cohesion.* The impact of shock and surprise can be seen to be a consequence of rapid, almost chaotic changes of the situation, or rather one's perception of it.

The Suez campaign of 1967 is a good example. The Israelis showed 'what can be achieved by surprise, determination and speed'.[29] It is not so much that they won the campaign, and very quickly, but that by their dynamism and use of the unexpected they broke the will of the Egyptian command, which resulted in the Egyptian army being ordered to break contact and withdraw rapidly. The result was little short of a rout, as photos of burnt-out columns in the Mitla and Jiddi Passes show.[30]

Secondly, the statistics illustrate once more that we tend to judge by what we can most readily measure. Men have talked about force ratios for centuries, yet only recently has a study looked at the impact of apparent intangibles and shown statistically that they have more effect than any likely range of force ratios. Combat is not as we have seen it. Battles are not about superior force, although that has some impact. Battles are about unhinging and disconcerting the enemy. It seems that they always have been. It is just that we have been unable to see that, because we have only recently been able to make such statistical comparisons.

Why are surprise and aggressive reconnaissance so effective? According to complexity theory, 'the largest amount of information comes from the least likely and most unexpected messages and outcomes'.[31] This is probably dichotomous. To the attacker, handling his reconnaissance forces aggressively, most of 'the least likely and most unexpected messages and outcomes' are likely to be beneficial. They will relate to the sort of things he wants to know, such as: where the enemy is; where he is not; and where there are gaps or weaknesses. Conversely, to the defender who is surprised, information about 'the least likely and most unexpected messages and outcomes' is unpleasant. The enemy has found a gap! Enemy reconnaissance troops are roaming your rear areas! A flood of such information will work like a shot of adrenalin. It might, if sufficiently small and if the defender is competent, spur him to rectify the situation. Conversely, if it is an overdose and the defender is not so competent, it might prompt panic or flight.

At what point is combat resolved? It is not when all individual one-to-one fights are resolved. Rarely if ever is all of one side killed, incapacitated or made prisoner. On reflection we see that the normal condition for tactical success or defeat is the collective withdrawal of participation.[32]

Not only is it more normal, it also appears to be easier to achieve. At some stage one side (or very occasionally both) consider that they are defeated, or must withdraw from combat before they are. That results in retreat or surrender. Achieving this seems to be easier than killing, incapacitating or capturing every single enemy; and sometimes markedly so. The Fall of France in 1940 is an example. The Germans crossed the Meuse at Sedan on 13 May 1940, and although the campaign went on for a total of six weeks the French Prime Minister Paul Reynaud thought that the campaign was lost by 15 May.[33] The Germans lost about 43,000 dead in the campaign, compared with almost 1.5 million dead to *fail* to conquer France in the First World War.[34] Similarly the Israelis attacked in the Sinai on 5 June 1967; the Egyptian high command panicked and ran by the late morning of the 7th. The Israelis completed the conquest of the Sinai in 89 hours.[35] There are several other examples.

Not only is collective withdrawal of participation more common than numerical attrition; it can also be decisive even if partial. If a part of an enemy force surrenders or withdraws, the whole of the attacking force can concentrate on those continuing to resist. This increases the effective force ratio in their favour, and probably also their freedom of manoeuvre. Thus seeking to persuade, coerce or force the enemy to desist is extremely attractive. It is a major element of the concept of the Manoeuvrist Approach.[36] Persuading or encouraging the enemy to desist from combat may occur in several ways. Surprise seems to be one of the most effective. This suggests that human will and participation in battle is a far more worthwhile target than we expect.

In general, defeat occurs when the enemy believes he is beaten. Marshal Foch considered that 'a battle is lost when one thinks he has lost, for a battle cannot be lost physically'.[37] This applies to a platoon commander as much as an army group commander. It suggests a mechanism for tactical success: persuade, coerce or force a platoon commander to believe that he is beaten, and in a real sense he is. Then use that event to persuade, coerce or force his company commander to believe he is beaten; then his battalion commander; and so on. Alternatively persuade the army group commander that he is beaten and his subordinates will tend to believe that they are, as well. The general point remains the same: defeat occurs when the enemy, at whatever level you are considering, believes that he is beaten. Defeat is a psychological state.

Defeat is thus largely a human phenomenon, and to a great extent a matter of belief. That raises an intriguing set of circumstances. Consider a fairly simple situation where one force attacks another. What is the actual outcome? There are nine possibilities, as shown in Figure 3.

Only in Cases 3 and 7 is there any categoric outcome. In all other cases the result is at least ambiguous. This says much about the uncertainty and fog of war, not least since knowing what the enemy commander believes is never likely to be easy.

Combat is a highly dynamic, complex, lethal interaction between human organizations (which are themselves internally complex). 'Between' is a key aspect. Combat is adversarial. Cautioning Moltke the Elder over flights of sweeping manoeuvre in the introduction to the latter's *Cannae*, General von Freytag-Lohringhoven observed that in the actual execution of grand schemes such as 'penetration of the front' and 'Reconnaissance, Victory and Exploitation', success is 'greatly dependent on the effects of the weapons of the enemy'.[38] We have also seen that a lack of consideration of the adversarial nature of combat is a criticism of the current British paradigm. Put simply, 'the other guy gets a go, too'.

Complex situations are difficult to understand. It is even harder to know what to *do* at any point, because determinism is weak: cause does not lead at all strongly to effect. That is not to suggest that humans cannot operate in that environment. Clearly they can. To do so they need to be able to form a high-level understanding of the situation. They need to be able to make inferences rather than deductions, and they need to accept that rapid, chaotic changes to the situation do occur, although not always to one's advantage. Soldiers need to

Attacker believes he has won:	1: Impasse	2: Uncertainty	3: Attacker has won
Attacker believes he has neither won nor lost:	4: Uncertainty	5: Uncertainty	6: Uncertainty
Attacker believes he has lost:	7: Defender has won	8: Uncertainty	9: Impasse
	Defender believes he has won	Defender believes he has neither won nor lost	Defender believes he has lost

Figure 3: Possible Outcomes

operate in that environment, and exploit its characteristics.

Thus to some extent what is needed is the skill of abstraction or simplification. Note that 'simple' is not the same as 'easy'. Things may be complex but easy, simple but difficult, or any other combination. Arguably it is training which allows people to perform difficult tasks with ease, whilst education allows them mentally to reduce complex problems to simple ones.[39] Note also that simple and simplistic are greatly different: simplistic is *unwarrantably* simplified.

The military historian Brent Nosworthy remarked that, in essence, tactics can be reduced to four elements: methods of organizing troops on the battlefield; how troops are moved; the most effective use of available weaponry; and the ability to react efficaciously to situations encountered on the battlefield.[40] This is a relatively simple statement. However, in terms of nested sets of systems, those relations will differ to some extent at every level from private to army group. So it may be true, and it might appear simple. That does not make it easy.

Combat is adversarial and destructive. It is an interaction that usually causes losses to both sides, although generally not in equal amount. To some extent this may be advantageous. Strategies of attrition are currently not fashionable because they are seen very narrowly as the opposite of manoeuvre.[41] This is another example of the relatively poorly developed state of military thinking. Objectively attrition is a sound tactic, as long as it is markedly one-sided.[42] The initial German tactics at Verdun appear to have been to inflict one-sided attrition on the French. At first it succeeded. German commanders then lost sight of that goal, and tactical conditions changed to the point where their initial tactics could no longer work.[43]

Destruction has another systemic effect. Considerable advantage accrues from the sudden destruction of a significant proportion of the enemy's force. Firstly, the remainder is likely to be thoroughly demoralized. Secondly, the force ratio (and probably also the force density) will change suddenly, to the advantage of the force which created the destruction. That may create opportunities for manoeuvre. There are thus two reasons why we should not dismiss destruction, simply because it is currently out of fashion.

Since combat tends to some extent to be damaging to both parties, some tactics seem appear advantageous. One is that one should, given a choice, be prepared to drive a long way rather than fight. A successful battle of manoeuvre is likely to be less costly than one with much fighting. Secondly, we should avoid iteration. If we take losses every time we fight, we should fight as infrequently as we can, and then with the maximum advantage on our side. We should, where possible, emulate Soviet theory and create the initial conditions of victory before the first shot is fired. We should certainly avoid any suggestion of 'going around the OODA Loop'; whether faster or slower than the enemy. We really should not want to iterate combat. We should create the biggest possible advantage, then impose the maximum damage, in one blow if possible. That blow should destroy such a portion of the enemy's combat power that he is suddenly and demoralizingly unable to persist in combat. And that applies at any level.

War is adversarial, and this also has an effect over time. Nosworthy remarked that in the Napoleonic Wars one tactical or grand tactical system could create and advantage for a year or so, or at most a decade.[44] Put differently, the more successful you are in war, the less likely you are to succeed the same way next time. This appears to be another example of paradox. It is perhaps best encapsulated as 'the loser learns most from war' – which might be called Williams' Law.[45] The reason is reasonably self-evident. The loser has just been beaten. He knows that if he does nothing, he has every chance of being beaten again the same way. He has a real incentive to overcome the advantage which defeated him. That advantage may be one of weapons, methods or training.

Examples abound. The Swiss dominated European warfare for less than a century before the Germans, notably Georg Frundsberg, raised their own pike formations – the Landsknechts.[46] Frederick the Great so thoroughly humiliated the French army in the Seven Years' War (1756–63) that it spent the next 20 or so years thinking through and refining its weapons, tactics and training. Guibert in tactics,[47] Gribeauval in the field of artillery, and even Baron Larey in the field of medical services, were all part of this movement. When the French Revolutionary armies took to the field after 1789, they had a solid base of modern equipment and technique.[48]

The tank makes a more complex example. It is simplistic to say that the tank broke the deadlock of trench warfare in the First World War; that the Germans then developed tanks in secrecy during the interwar period; and the British did not. However there is a germ of truth in that. The eventual British solution to trench deadlock was a tactical method built around tanks. The German solution to trench deadlock was stormtroop tactics, based on hurricane bombardments and aggressive infantry infiltration on multiple routes. After the war the Germans realized that the wider, operational-level reason why they had almost won the war in the West in 1914, but eventually failed, was a lack of mobility.[49]

They saw that the tank could provide that mobility. The problem was pressing for them, since they had failed to achieve their war aims in 1914–18. To that extent, they had lost. They performed secret experiments in the interwar years, much heralded afterwards. The main spur to development came from observing the British. An Experimental Mechanized Force was established on Salisbury Plain in 1927. Its activities were freely reported in the open press – Liddell Hart was the War Correspondent to the *Daily Telegraph*.[50] The Germans read those accounts and translated them for domestic military consumption. They felt that the British were ahead of them in tank force development.[51] At the time, they probably were. The British initiative petered out, not least due to the appointment of the somewhat reactionary Field Marshal Montgomery-Massingberd as Chief of the Imperial General Staff in 1935.[52] Conversely the German initiative did not. The Germans raised their first panzer division in 1935 and deployed ten in the French campaign. The British did not field a complete armoured division until 1940 – and that was in North Africa, not France.[53]

Broadly speaking, the British had created an advantage with tanks in the First World War, but the Germans had completely overturned it before the Second World War. The Germans created an advantage of their own by deploying integrated all-arms armoured formations. Yet the world turned again. By about 1943 the Allies had negated that advantage. At lower levels they deployed efficient anti-tank weapons and improved training. At higher levels they fielded their own armoured divisions.

Armoured warfare provides a further example. In 1956 and 1967 the Israelis defeated numerically superior Arab forces through superior use of armour and CAS. The Arabs responded by deploying massed anti-tank guided weapons (ATGW) and anti-aircraft systems in the Yom Kippur War of 1973. Where the Egyptians forced the Israelis into early armoured counterattacks on the Sinai front, that had considerable success. Similarly, where the Israeli air force tried to support the army with CAS missions, it ran into difficulties and took significant casualties.

That is a good example of the loser learning most, but the sequel is illuminating. Within the 18 days' duration of the war, the Israelis reorganized their all-arms mix and their tactics.[54] That largely overcame the Arabs' technical advantage. The Israelis counterattacked across the Suez, surrounded the Egyptian Third

Army and forced Sadat to sue for peace.[55] The Israelis received TOW ATGW from the USA during the conflict, and also defeated the Egyptian anti-aircraft 'umbrella'. Methods included improved tactics and the use of hastily imported US countermeasures such as chaff, flares and jammers.

Thus war is evolutionary – as Clausewitz and others have remarked.[56] Overcoming an enemy's advantage can be done in several ways. One is imitation – such as the Landsknecht, or deploying TOW. Another is technical counters – ATGWs, surface-to-air missiles (SAMs), chaff and flares. A third is tactical and organizational developments – French column tactics after Guibert, or armoured divisions. Some of these counters can be very fast – within days, if needs be. In practice, however, the really effective and enduring developments are those which are thought out over long periods of peace. They typically also provide a lasting change to the nature of warfare. War is evolutionary, and that allows original and novel thought. That is the gateway to creativity for the practitioner. Warfare may be neither art nor science, but indisputably has a creative element. And since war is adversarial, you can never fight the same enemy twice in the same way – unless he is stupid (in which case you probably won't have to). In practice, no two wars are ever the same.

We have tried to develop a necessary and sufficient paradigm or description of combat. It would be difficult to say whether we have achieved that. Combat is adversarial, human, destructive and evolutionary. Moreover combat is a dynamic, complex interaction between organizations that are themselves complex. Critically, human aspects such as shock, surprise and offensive reconnaissance are more effective than force ratios. The human effects are all-important. People matter most.[57]

We should base our understanding of combat on a fundamental premise: that fighting battles is basically an assault on the enemy army as a human institution. Combat is an interaction between human organizations. It is adversarial, highly dynamic, complex and lethal. It is grounded in individual and collective human behaviour, and fought between organizations that are themselves complex. It is not determined, hence uncertain, and evolutionary. Critically, and to an extent which we currently overlook, combat is *fundamentally* a human activity. This, then, is our new paradigm.

> The evidence is overwhelming that behavioural considerations – such as combat effectiveness, leadership, and surprise – were considerably more important in 1973 than a purely material comparison of men, numbers, weapons and technology.
>
> If the October War proved anything, it demonstrated that the human element in war remains as important as it ever was.[58]

The above quotations refer to the Yom Kippur War of 1973. Similar observations relate to almost every conflict. At the time of writing, the most recent warfighting campaign which the British Army had analysed was the period of major combat operations in Iraq in 2003. Its analysis found that 'it also appears that the contemporary operating environment is yet more complex and chaotic (hence confusing) than perhaps previously thought'.[59]

We have probably not yet developed a full paradigm of combat, in all the richness required. To do that would require a detailed knowledge of all the emergent properties of armed forces, or all the emergent characteristics of combat. We will explore some of these later. However, we will never reach a full description. That is inherent in the problem. We should not expect 'the answer'. There is probably no 'theory of war' in a form which we can easily recognize. War is evolutionary and dynamic: it changes all the time. Change is part of the description of war. It should be present in any description of warfare.[60] What we can expect, and should work towards, are ever-better views of the problem and useful solutions to it.

Little in this chapter should be particularly surprising. New paradigms aren't generally like that.[61] Cathedrals didn't fall down when Newton developed his theories of physics. Conversely, after Newton we could build the Eiffel Tower, the Golden Gate Bridge and the Sydney Opera House. The world didn't spin off its axis when Einstein discovered relativity; but since then we have been able to develop nuclear power, and detect quasars and black holes.[62] A quote from Marshal Marmont, writing almost two centuries ago, is illuminating:

> Tactical talent consists in causing the unexpected arrival, upon the most accessible and important positions, of means which destroy the equilibrium, and give the victory; to execute, in a word, with promptness, movements which disconcert the enemy, and for which he is entirely unprepared.[63]

That is to say, one should attack the enemy's will with speed and surprise. There is, in the end, nothing new under the sun.

4

Tools and Models

Paradigms don't win wars. It might be said that the revolutionary French armies or the Wehrmacht won battles or campaigns because they fought according to what was at the time a superior paradigm. You don't, however, see a paradigm on a battlefield. You see organizations and tactics, which reflect the doctrine and the training which an army employs. At a high level, organizations and training reflect big ideas – concepts – about what wars and battles are like and how they should be fought. We need to develop a set of concepts based on our paradigm. Those concepts should be fairly concrete: they should mean something to the practitioner. Our first requirement, however, is for intellectual tools to help us move on from our somewhat abstract paradigm.

Rationality and logic serve us well for some aspects of warfare. They tend to be the more technical aspects such as artillery, engineering and logistics. But there is a down-side, which is a reliance on a mechanical paradigm, and hence, a reliance on weapons effects.[1] Combat requires a different approach: one that is equally logical but starts from a different perspective. That perspective is holistic, empirical and pragmatic. It also acknowledges the human dimension, for 'only imaginative, empathic insights, rather than abstract and universal principles, could penetrate into the wealth and uniqueness of human reality'.[2]

We shall look at two simple models and one piece of theory. Clausewitz's 'Dialectic of aims and means' helps us understand the wider implications of the adversarial nature of combat. General Systems Theory (GST) allows us to consider armed forces as complex entities. Archer Jones's model of troop types gives us some insight into how various types of forces interact on the battlefield.

Clausewitz suggested that the aims of a party to a conflict should be proportionate to the means at hand. He also suggested that the aims of both protagonists interact, as do their means.[3] Clausewitz described this process as a 'dialectic of aims and means'. Clausewitz's term *Zweck* literally means 'purpose', here 'aim'. An alternative sense is 'end', as in 'ends and means'.[4] However, the meaning is highly contextual; 'aims' works better in the present sense. For the present discussion we shall take 'means' to imply both 'the means *with* which' as well as 'the means *by* which'.

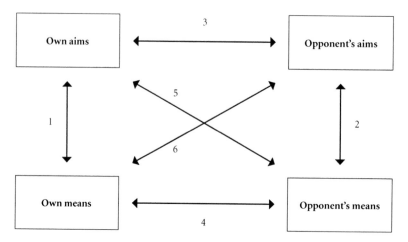

Figure 4: Six Interactions

Thus a 'dialectic of aims of means' implies an interaction of intentions, forces and methods. The dialectic describes six possible interactions, as in Figure 4.

The aims of a party to a conflict should be proportionate to the means at hand: hence Relationships 1 and 2. War, and combat, is adversarial. Thus thwarting the enemy's aims should be a factor in one's own aims, and vice versa. Hence Relationship 3. Furthermore, committing the means to combat affects their nature (not least through attrition). Hence Relationship 4. The nature and capabilities of the enemy's forces is a factor in considering one's aims. Hence Relationships 5 and 6.

This two-dimensional representation of aims is reasonably comprehensible. To distinguish 'aims and means' into 'ways, ends and means' risks overcomplicating an already complex set of relationships, and is not attempted here. I do not believe that Clausewitz ever used it. 'The dialectic of aims and means' is peculiarly succinct. As we shall see, it is also a powerful intellectual tool.

The immediate and obvious observation is that both the enemy's forces and his intentions should be factors in considering courses of action. There are two further deductions. Firstly, possessing and retaining a fit instrument is a necessary condition to being able to take part. Without means, one cannot fight. The state of the instrument is highly important. It should be numerically strong, well equipped, well trained and well administered. But it must also be protected against the enemy's actions, and those of the environment. Protection is absolutely essential. This is a major consequence of the adversarial nature of combat. The second deduction is the importance of destruction. If part of his means are destroyed, the protagonist is less capable of achieving his own aims, less capable of affecting the enemy's means and has less impact on his aims. Destruction is a key aspect of conflict.

Destruction was obviously and quite explicitly denigrated in the British military doctrine of the 1990s.[5] This should be seen in its historical context. The advent of explicit doctrine, started by Field Marshal Sir Nigel Bagnall in the late 1980s, sought to establish a manoeuvrist approach and play down the role of attrition. Thus the attack on will and cohesion, at least implicitly involving physical manoeuvre, was elevated to the denigration of destruction. The relevant pamphlet stated that 'The costs and potential aims of the enemy's destruction must be weighed realistically and not overemphasized.'[6] This is a valid stance, but ran the risk that a generation of officers would forget the importance of destruction *per se*. We will discuss destruction and its corollary, protection, later.

The Dialectic is intended to be a descriptive tool. It is not intended to have moral or ethical overtones. We considered the moral aspects of pragmatism earlier. The Dialectic can be loosely translated as 'the end justifies the means', which enables a moral interpretation of the word 'justification'. Whether or not that has any relevance to Clausewitz is not discussed here. We are considering the conceptual not the moral component of fighting power. No moral interpretation is intended.

In the 1950s von Bertallanfy sought to describe complex biological systems holistically, with features that are common across different systems. He described a system as containing a series of interlinked elements.[7] The system performs some form of transformation process, and is contained within a boundary. As an example, a jam and pickles factory can be seen as a system. The elements might be the machines; the linkages might be conveyor belts; the transformation process would be changing raw fruit, sugar and water into jam and pickles. Traditionally the boundary would be a brick wall around the factory. Figure 5 describes the basic elements of such a system.

The system has inputs and outputs. In this case the obvious inputs are the ingredients. Less tangible inputs include electrical power, human labour and even the injection of capital. The outputs are tins or bottles of jam and pickles. Useful systems also have outcomes, which are the consequences of the output on the external environment. In our case the intended outcome might be the lifestyle of the owners of the factory, the lifestyle of the employees, or employment in Third World villages. In real, complex systems there could be several outcomes, not all of them desirable.

When we consider useful systems we can also consider economy, efficiency and effectiveness – the Three 'E's – as represented in Figure 6.

In systems terms, 'economy' relates to minimizing inputs. 'Efficiency' relates to the ratio of inputs to outputs. 'Effectiveness' concerns the maximizing of outcomes. The three terms have no formal relationship to each other. For example, it is quite possible to economize and be less effective. One could put less fruit in the jam. It

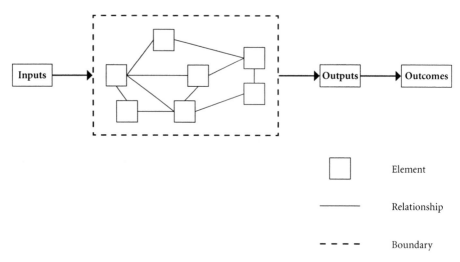

Figure 5: A System

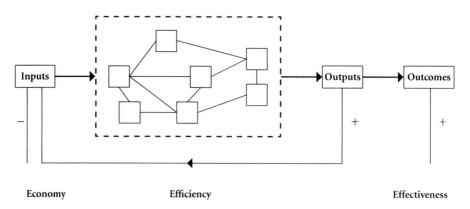

Figure 6: Economy, Efficiency and Effectiveness

would cost less, but we might sell less and our profits might go down. Conversely it is possible that reducing the fruit content and retail price of the jam would place the product in a different market sector, and profitability would increase.

Profitability might increase or decrease for other reasons, because the jam factory is only one component in a wider system: the economy. Outcomes are the consequences of the output on the external environment. The external environment will normally be a wider and more complex system.

We have previously discussed the idea of units and formations being nested systems. Nested systems are an accepted aspect of GST. One function of the system boundary is to identify the system as a component, or element, within a wider system. The boundary also helps to identify inputs into, and outputs from, the system. In our factory the inputs were readily identifiable, not least because of the existence of a factory wall. Inputs are in the first instance things delivered to it, largely by truck. They cross the boundary: they go through the factory gate. Power and labour are exceptions (they don't arrive by truck). They are fairly easy to identify, as are outputs. In the wider application of GST, inputs and outputs are harder to identify. So are boundaries, relationships and components. Nevertheless GST is useful, not least because it forms a taxonomy. We previously asked whether the Forward Observation Officer forms part of the battlegroup he supports, or his parent artillery unit. He provides an input to the battlegroup, fire support, which is an output from the artillery unit. He also provides tactical information from the battlegroup to the artillery unit. Thus he provides inputs to both systems. He has direct relationships to elements of both systems, so should be considered part of both.

Complex systems also have emergent properties. These are characteristics of the system which are not predictable in advance. For the jam and pickles factory these might be the smell and the noise detectable outside the factory. The term 'predictable in advance' should be applied with caution. It might be obvious to some people that assembling machines to turn fruit, sugar and water into jam would be smelly and noisy, but not to others. The expression 'not in practice predicted in advance' might be more useful. Organizational culture is a typical emergent characteristic of a human organization.

GST allows some understanding of complexity, and complex systems, through a holistic and global approach. Specifically, instead of trying to decompose difficult and complex relationships between elements, we can consider the system as a whole. We can describe (and sometimes even predict) the behaviour of that system. If we can get towards the point of predicting the behaviour of large, complex systems such as divisions and brigades we can say much about the nature of combat and how to conduct it effectively.

We have seen that war generally, and combat specifically, is dominated by considerations of effectiveness. Having beneficial outcomes – winning – is all-important. Armies do not get paid to come second. The means and the method are relatively unimportant, except in so far as they contribute to achieving beneficial outcomes. We can make several observations.

First, outcomes are the consequences of outputs on the external environment. Thus even if the means performs perfectly in relation to the aims, and the aims are perfectly formulated, the outcomes are not always predictable or desirable. The archetype of such systemic behaviour is probably the army of Frederick the Great. He inherited (and attempted to perfect) what was seen at the time as an almost perfect military machine.[8] That is far from saying that Frederick won all his campaigns, let alone his wars.

More generally, complex systems can have rogue outcomes. Some undesirable outcomes are reasonably obvious and can be guarded against. For example, the use of unlawful or immoral means by one's own troops may well raise a political backlash which renders the purely military outcomes largely irrelevant. Western armies go to great lengths to guard against such practices, not just for such cynical reasons. Unfortunately rogue outcomes are, by definition, unpredictable. War, and particularly combat, entails risk. Handling risk is a major issue in combat. The incidence of rogue outcomes is an example of the law of unintended consequences, which is particularly relevant to complex human situations. It can be stated flippantly as 'let no good deed go unpunished'.

Importantly, rogue outcomes may be beyond any one commander's power to redress. This seems fundamental to war. Since rogue outcomes may occur no matter what precautions are taken, methodological precautions are not enough. The idea of rogue outcomes is well entrenched in military culture. Examples are Murphy's Law (that if something can go wrong, it will go wrong; and often already has); Sod's Law (that Murphy was an optimist) and, deeply in the vernacular, 'sh*t happens'. The flexibility to respond to rogue outcomes as they arise is far more important than methodological precautions that attempt to prevent them.[9] This requires three things:

- Vigilance, to detect rogue outcomes at the earliest moment.
- Mental and methodological flexibility, to be able to react in ways that were not predicted in advance, and
- Redundancy. This requires the provision, retention and if necessary regeneration of reserves. This means true reserves, not echelon forces.

Reserves are, strictly, uneconomical. In a perfect world they are not used, and thus in terms of outputs gained against assets deployed they are an unwanted overhead. War and combat are not, however, perfect. Our paradigm of combat is that of a messy, unstructured, highly imperfect but distressingly real environment. As van Creveld put it, in war efficiency and effectiveness are not complementary, but opposed.[10] That is, to be effective we must be inefficient. We should strive for effectiveness, not quality of process. For many this will be a considerable change of view, akin to the leap from seeing epidemiology not as the study of the progress of disease through a population, but the study of staying healthy.[11]

Economy and efficiency are not the same as effectiveness, and this is particularly true in combat. I once saw film footage of a US attack aircraft attacking a sampan in Vietnam. The pilot fired into the water below the boat to see the fall of shot. Then, as he flew towards the boat, he 'walked' the fall of shot onto the target in one long burst lasting several seconds. The sampan virtually disintegrated into (to make a bad pun) a hail of junk. The attack was highly (and obviously) effective: the boat was completely destroyed. However, several hundred rounds were fired, and only a few dozen struck the boat. Therefore it was at first sight neither economical nor efficient. Conversely, and on another level, to the USAF it was probably quite efficient: one sortie, one boat, for a few hundred dollars' worth of ammunition fired. Thus although descriptions using the Three 'E's are useful, it is important to use them in the correct context.

Van Creveld observed that extreme efficiency in human organizations normally results from very close internal coordination. That is, effectively, the streamlining of internal process. However, such close coordination is vulnerable to disruption and typically inflexible.[12] Military units need to be robust to survive. Hence effectiveness requires units and formations to be inefficient *and* uneconomical, to some extent.

The American historian Archer Jones reviewed over 4,500 years of military history and concluded that there are fundamentally four basic troop types, the nature of which and the relations between which pervade over time.[13] He discriminated between more- and less-mobile troops, and between troops which act by missile fire or by close combat. These two determinants define his four basic troop types. Figure 7 shows the model of troop types c. AD 1200.

In order to make the model more widely applicable, we need to be able to describe 'more mobile' troops that act through close combat without referring to horses. We shall use the term 'Shock Troops' for 'Heavy Cavalry'. Furthermore, although light cavalry (such as the Huns, Mongols and Saracens) did use bows, their effectiveness relied more on their raiding and skirmishing tactics

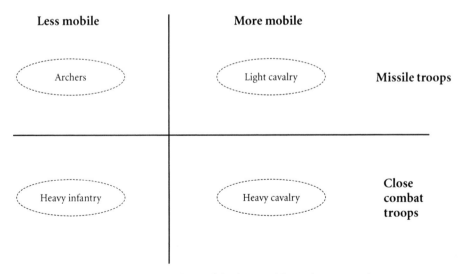

Figure 7: Generic Model of Troop Types (c. AD 1200)

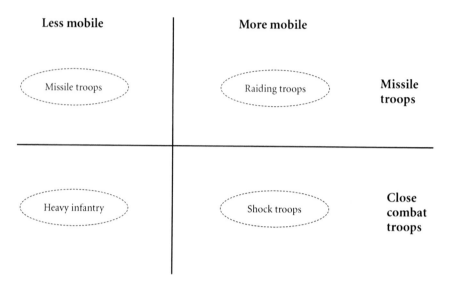

Less mobile | **More mobile**

Missile troops

Raiding troops

Missile troops

Heavy infantry

Shock troops

Close combat troops

Figure 8: More Generic Model of Troop Types

than their missiles. Thus for the present discussion Figure 7 can be redrawn as Figure 8.

Figure 8 is only a classification. The real value of the Archer Jones model comes from the relationships between those troop types. For example, he considered that well-ordered heavy infantry can normally resist the attack of heavy cavalry. The Battle of Hastings demonstrates the point: until disordered by archers, King Harold Godwinsson's men might have won the day.[14] The Battle of Bannockburn and some of the Swiss victories in the fourteenth and fifteenth centuries reinforce the point.[15] Their physical and social cohesion appears to have granted the infantry protection above that of their armour alone. Similarly, well-ordered missile troops can resist the attack of raiding troops. There are few good examples of this, but Richard the Lionheart's march on Jerusalem in 1191, in which he used his crossbowmen to keep the Saracen horse archers at a safe distance, is one.[16]

Missile troops can normally prevail against heavy infantry, being sufficiently nimble to avoid being attacked, and yet having the lethality to damage them. The *coup de grâce* delivered by Xerxes' archers against the Spartans at Thermopylae is a good example. Perhaps surprisingly, missile troops can also prevail over shock troops. Their firepower is in practice sufficient to prevent shock troops from closing, whilst at the same time inflicting critical damage. The classic examples are probably the major battles of the Hundred Years' War, in which French cavalry repeatedly failed to defeat the English archers.

Similarly, raiding troops can normally prevail over shock troops; classically, light cavalry over heavy. Raiding troops can harry, harass and tire the shock troops until the latter are exhausted and lose their effectiveness. There are not many good examples. The Battle of the Horns of Hattin in AD 1187 during the Crusades might be one. Finally, raiding troops can normally defeat heavy infantry. Their mobility and their ability to damage at a distance eventually tell over the protection of the heavy infantry. Crassus' defeat by the Parthians at Carrhae in 53 BC is an example.[17] The six relationships can therefore be summarized by Figure 9.

These relationships *are* generalizations: they allow exceptions. The major categories of exception are tactics, terrain and technology. Richard's march on Jerusalem was an exception of superior tactics. His use of missile troops allowed him to overcome the traditional advantage of raiding troops over his knights, his shock troops. His tactics were a *combination* of troop types used in synergy. The German Emperor Otto's victory over the Magyars on the Lechfeld in AD 955 was an example of the use of terrain. Otto managed to pin the Magyar light cavalry against the River Lech. Unable to manoeuvre and so exploit their innate advantages, the Magyar 'raiding troops' were decimated by the German heavy cavalry.

Technology further confuses the picture. Where a force has protection that simply cannot be overmatched by their opponents' weapons, or they have mobility that enables them to avoid damage in most circumstances, Archer Jones's pervading relationships will not apply. However, technology is neutral and does not stand still. The tank – anti-tank 'duel' in the Western Desert of 1941–43 is

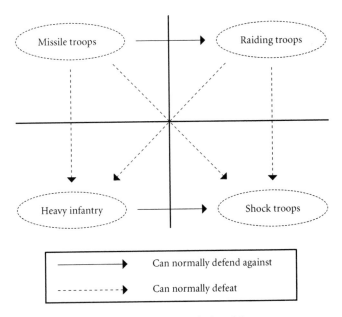

Figure 9: Generic Relationships

a good illustration. Taking tanks as 'shock troops' and anti-tank guns as missile troops, in a relatively straight fight the anti-tank guns should have prevailed. However, where the tank was a German Panzer III or IV and the gun a British 2-pounder, the tank won on a simple issue of weapons effectiveness. Once the British had 6-pounders or even 17-pounders, the balance was redressed.[18]

Even at this early stage Archer Jones's model has shortcomings. It does, nonetheless, give insight and is worth persisting with. What emerges from Jones's work is a perception that, without some mechanism of shock action, combat is rarely quick nor particularly decisive. Whatever shock action is, and Jones scarcely uses the term, it denotes close combat that is somehow different from the simple application of missile power. That finding, which summarizes about 700 pages of narrative and analysis, is critically important. *Without some mechanism of shock action, combat is rarely decisive.*

Jones's findings indicates that missile power may eventually prevail, largely through numerical attrition, but that in practice it rarely does. That explains why instances of missile and raiding troops defeating each other or close-combat troops are few. They might prevail, but in practice they rarely do in any quick or decisive way. The Parthians needed two days to destroy Crassus' army at Carrhae. Saladin also needed two days to destroy the Crusader army at Hattin.[19] Thus there must be some mechanism of shock action for combat to be sudden and decisive. It is therefore highly important to understand what shock is, how it is applied and what its effects are. We will look at that later.

We should make a further observation about Archer Jones's model. We described superior tactics as a superior combination of troop types. Jones makes this quite explicit: superior tactics are, effectively, making use of your own troop types in a better combination, in order to negate the enemy's strengths and exploit his weaknesses.[20]

As soon as Jones's model is expanded beyond four troop types the inter-relationships become numerous and messy. Some proliferation is nonetheless necessary in order to update the model into the twentieth century. Figure 10 relates to 1944–45.

The high mobility and largely transient effects of aircraft makes them the modern raiding troops. Field artillery has replaced archers as missile troops. Tanks are the modern shock troops. Infantry, given time to dig in, is the 'heavy infantry' of old. However, by 1944 or so the pervading relationships had been skewed by technology. Protection had given the tank a superiority over infantry in some circumstances. The sheer speed of aircraft, coupled with the use of the third dimension, enabled them to roam over unprotected gun lines at will. The solution to both problems was the addition of specialist missile troops.

Anti-tank artillery could, with occasional swings of the technological pendulum, defeat tanks and therefore restore effective protection to the infantry.

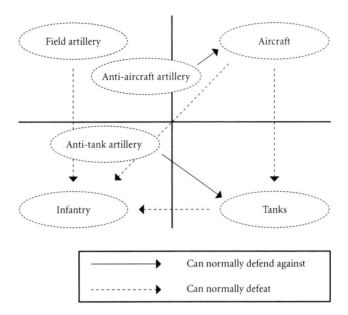

Figure 10: Relationships c. 1944–45

Critically it was their missile effect, and not the ability to engage in close combat, that bestowed this benefit. The battle of Medenine in 1943 was a good example. The Germans lost 52 tanks on the first day, of which 45 to anti-tank guns.[21] Similarly anti-aircraft artillery, once deployed, could generally protect ground troops in its vicinity from aircraft.

This taxonomy allows us to comment further on airpower. Airpower enthusiasts may take exception to the description of airpower (or at least Offensive Air Support [OAS]) as 'raiding troops'. However, whilst aircraft can sometimes be highly effective, their presence is transient. The planes fly away. In Normandy, large numbers of fighter-bombers seriously affected the movement of German forces in daylight.[22] Aircraft have traditionally excelled at reconnaissance and interdiction, not least in denying the enemy the use of his long-range sensor systems. Yet even in the Gulf War OAS only accounted for about 20–25 per cent of the attrition of Iraqi armour.[23] Interdiction of supply lines did not have a critical effect;[24] and airpower did not degrade Iraqi theatre-wide communications to a critical level.[25] Even on the Basra Highway only 28 AFVs were destroyed.[26] That reflects eerily on statistics from the Normandy campaign in which Allied Thunderbolts and Typhoons destroyed only one German tank for every ten soft-skinned vehicles.[27]

More recently, in the Kosovo air campaign of 1999 aircraft were kept at arm's length (above 15,000 feet; at which range weapons targeting was extremely difficult) mostly by the threat of AAA and shoulder-launched SAMs, which are easy to conceal and very difficult to counter. In Normandy, the Gulf and Kosovo the air battle went on for weeks, and the campaign was only concluded through ground manoeuvre, or its threat. Airpower made significant numerical impact in the first two campaigns once the enemy was broken. Thus we have a troop type which is highly mobile; relies primarily on missile power; is of transient effect; is particularly useful for raiding and scouting; and is highly effective at harrying a broken enemy. It can, however, be kept at bay by a well-ordered enemy using missile power. Frederick would have asked no more of von Zeiten, his cavalry commander, nor Napoleon of Murat.

This discussion of airpower, although relevant in itself, highlights a benefit of the Archer Jones model: taxonomy. The model allows discussion to be structured, and the development of a hypothesis about shock action. Without that taxonomy, much of the debate over the relative merits of airpower, for example, resembles Newton's observations about dung-heaps: interesting but irrelevant, and rarely insightful.

We have reached the limits of the Archer Jones model. Its troop types and relationships are generalizations. That is respectable. War is not determined, and only the pedant should expect relationships that never admit of exception. The major classes of exception (tactics, terrain and technology) are useful. Once we try to go beyond four basic troop types the relationships become numerous and hard to understand. Thus, as a model, it is limited but useful.

Clausewitz's Dialectic has a major implication: the importance of destruction. Consider Figure 4 with a significant part of the enemy's forces destroyed, as shown in Figure 11.

The impact is significant. The enemy's ability to prosecute his aims is reduced; his forces' ability to damage and otherwise impede one's own forces is constrained; and one's own freedom of operation is considerably greater. Hence relationships 2, 4 and 5 are directly affected; relationships 1, 3 and 6 indirectly. This applies to any commander in any situation. Destruction has a major impact on the dynamics of combat.

We need to understand what 'destruction' implies. Clearly it must include some reduction of capability. It must be more than transient: surprise, shock and suppression are important but result in only transient loss of effectiveness. Thus destruction implies an appreciable reduction of capability for at least the duration of the current engagement. It may not be permanent. The impact of 'destruction' depends on the implications of 'the duration of the current engagement'.

A knocked-out tank might be considered 'destroyed' if knocked out and abandoned for that day's fighting, longer if it cannot be recovered. It might, however,

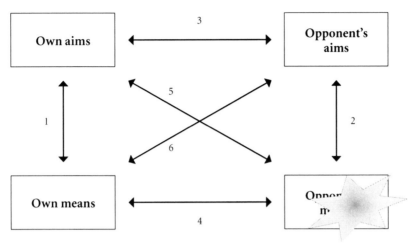

Figure 11: The Effect of Destruction

be repaired in hours or perhaps days. A tank unit is therefore quite difficult to destroy if it is provided with a supply of replacement tanks and crews. The unit might be sufficiently mauled in a battle to be rendered ineffective for that battle or that campaign. It might be reconstituted later. Whole formations can apparently be 'destroyed' but then reconstituted, probably after several weeks.

The Normandy campaign provides several good examples. Roughly 27 German divisions were considered to have been destroyed, mostly in the Falaise Pocket.[28] Yet at least eight were recreated within 12 weeks.[29] Two of those (the 271st and 344th Infantry Divisions) were dispatched to the Russian Front – scarcely a sensible option unless they were reasonably sound. Were they recreated, or were they fresh formations? The question is difficult to answer. Divisions answering those names entered combat in the stipulated timeframe. It is not reasonable to suggest that entirely new formations, having only the names of the predecessors, were raised in such a short period. Witness the fact that many of them subsequently fought well: these were not green formations. The Germans lost about 65,000 men killed and taken prisoner in the Falaise Pocket, but about 135,000 escaped and were available to fight again.[30]

The 21st Panzer Division, encountered by the British 3rd Division around Caen on D-Day, had been recreated after 'destruction' in Tunisia.[31] Of an initial cadre of about 3,000 men, some 2,000 had served in 21st Panzer before.[32] The other thousand were veterans from Africa, Crete and Russia.[33] Thus although 21st Panzer was 'destroyed' in Tunisia, an effective division with the same title and 3,000 veterans fought in Normandy a year later. It seems that where a viable cadre and command team survived, the division was recreated. Where no viable command team survived, a new division was raised.

We can nonetheless say that about 27 divisions were destroyed *for the purposes of the Normandy campaign.* As the Allied armies broke out across France, those divisions were not present to hinder the Allies' advance (except as PW and broken remnants fleeing eastwards). That is, they were *effectively* destroyed. Unsurprisingly, given what we now understand about the nature of war, it is difficult to be more specific.

Some aspects of destruction may seem categoric, but even that is misleading. A soldier who is killed can be considered to be 'destroyed', as can a tank that is so badly damaged that it cannot be repaired. However, soldiers and tanks can generally be replaced. From a systemic viewpoint, and accepting the human tragedy thus entailed, even this destruction is transient.

Capture, a far less complete form of 'destruction', can have a similar overall effect. A soldier taken prisoner is effectively *hors de combat* for a long time. The 21,000 British taken prisoner on 21 March 1918, the first day of the German Kaiser's offensive, might have been replaced in a week or a month. Yet for the purpose of resisting the German army on 22 March they were just as ineffective as the 2,512 killed and perhaps 10,000 wounded the previous day.[34]

Destruction, therefore, means an appreciable reduction of capability for at least the duration of the current engagement. It is not an absolute effect when seen from an overall, systemic viewpoint. From the perspective of GST, it means that certain elements of the overall system are no longer present, and that the interactions between them and the remainder no longer function. The system cannot function as before, at least for the duration of the current engagement. Within this definition, capture is effectively a form of destruction.

Unreliable and often inflated claims of destruction are minor but persistent characteristics of war. In some circumstances this is simply a matter of reporting. It appears that claims of enemy aircraft shot down in aerial combat have often been inflated, even though in many cases this was done in good faith.[35] We can now see that there is another factor at work: the meaning of destruction. It was arguably reasonable for the Allies to claim to have destroyed 27 divisions in Normandy, in early September 1944. But that did not prevent the Wehrmacht from fielding eight of them three months later.

Propaganda is another factor. We have already discussed Soviet claims of destruction of German armour at Kursk, in which Soviet figures greatly exceed the number of tanks replaced or repaired by the Germans.[36] After the Second World War the Red Army claimed to have destroyed 507 Axis divisions, compared with 176 destroyed by the Western Allies,[37,38] for a combined total of 683. However, the Axis raised only 565 divisions.[39] The total of 565 divisions raised ignores those that were rebuilt and re-entered combat a few months after apparently being destroyed during the war. It would not count, for example, the eight formations from Normandy that reappeared that autumn. Counting all such occurrences of rebuilding divisions, the 565 Axis divisions actually raised might reach the total of 683 claimed by the Soviets.

Nevertheless, on 1 April 1945 Germany still fielded about 230 divisions.[40] Admittedly, most were greatly reduced in strength and some were no more than 'paper' formations. That all of them eventually surrendered *en masse* (many of them no doubt to the Red Army), tends to undermine the Soviet claim to have *destroyed* 507; unless one counts divisions that appear to have been intact only a week before the cessation of hostilities. About 80 Romanian, Italian and Hungarian divisions also surrendered *en masse* when their countries surrendered.[41] This would reduce the total of 683 allegedly destroyed by about 230 German and 80 other Axis Divisions to about 370. Yet the Red Army claimed to have destroyed 507? Clearly, at a formation level the meaning of destruction is not clear-cut.

As we have seen, the British doctrine of the 1990s denigrated destruction. This seems to have been a deliberate consequence of the adoption of a Manoeuvrist Approach, which supports the enemy's defeat through attacks on his will and cohesion rather than through attrition. On one level this is supportable: if the Manoeuvrist Approach aspires to avoid bloody frontal slogging matches and prefers more cost-effective approaches, it should be a good thing. Conversely it has two disadvantages. The first is that published doctrine did appear to close serving officers' minds to the place of destruction in the dynamics of combat, as we have seen. Negative remarks about destruction as being 'attritional' were commonplace. The fact that doctrine supported its employment in certain circumstances was often overlooked.

The second disadvantage is that attrition is not necessarily detrimental. It is only detrimental if the attrition is inflicted on one's own forces. A strategy of one-sided attrition is perfectly respectable, if it can be achieved. The German army seems to have employed it to good effect in the very early stages of Verdun.[42,43] The risk is that it will turn into *mutual* attrition, as indeed happened at Verdun. As we shall see, however, it is probably less effective in many circumstances than what we understand as a Manoeuvrist Approach, and also less efficient.

If destruction is an important part of the dynamics of combat then so too is protection, simply because war is adversarial. The benefits of destruction apply equally to the enemy (and humanitarian considerations are of course a further issue). The obvious components of protection are hardening, fortification, dispersion and concealment. 'Hardening' is used here to include aspects such as armouring vehicles and, where appropriate, men. Fortification is at first sight unfashionable (the fate of the Maginot Line and the Bar Lev Line saw to that), but infantry still digs in, and combat aircraft are provided with bunkers, under the smart title of 'hardened aircraft shelters'.

Dispersion has obvious benefits, but some interesting disadvantages. The need for dispersion imposed by the weight of artillery delivered in the middle years of the First World War made front lines susceptible to infiltration, as the Germans rapidly discovered.[44] The 21,000 British prisoners taken on 21 March 1918 were to some extent victims of dispersion. Dispersion also makes human interaction more difficult. For example, a platoon commander has difficulty visiting all of his platoon's trenches under fire. These disadvantageous aspects can be seen as emergent characteristics of combat. In 1915 commanders probably did not foresee that increasing the weight of artillery bombardments would lead to large numbers of prisoners being taken two to three years later.

There are other, less obvious, facets of protection. One is protective firepower. The image of US firebases in Vietnam resisting attack with curtains of direct and indirect fire is a powerful example.[45] There is evidence that the only occasions on which German infantry penetrated British positions in France in 1940 were where the defender's machine guns were sited to fire frontally, rather than to a flank.[46] Armies had learnt the value of interlocking curtains of machine gun fire during the First World War.[47]

We saw earlier how specialist, less mobile missile troops (anti-aircraft and anti-tank units) were added to armies for protection. The Wehrmacht appears to have conceived of tank-destroyer units as protecting the formation of which they were part:[48] an interestingly systemic view of armoured formations. Another interesting development is the speed with which English-speaking armies dropped anti-tank units after the end of the Second World War. We will return to that discussion later.

Another aspect of protection only appears when armed forces are viewed from a systemic viewpoint. Archer Jones charted the development of articulation from the unitary hordes of Bronze Age warriors, through the linear tactics of Marlborough and Frederick, to the section-level tactics used from 1917. He pointed out that highly articulated forces, which consist of a large number of discrete but linked parts, are harder to destroy *en masse* than unitary forces.[49] Once King Harold's shield wall was broken at Hastings the day would be lost, and Harold probably knew it. Conversely, breaking one square in a Wellingtonian battle, rare though it was, never resulted in the defeat of an army as a whole.[50]

There also appears to be a connection between morale and decentralization, and hence articulation. It seems that, within reason, commanders and staffs of decentralized groups tend to feel more empowered than those in more centralized organizations. They are better motivated, and feel more responsible for the outcomes of their actions.[51] This seems to support the apparent effectiveness of German battlegroups in the later stages of the Second World War. Equipped with their commander's intent and considerable delegated authority, they were both effective and resilient. Their ability to take in reinforcements and return to battle must have been a major factor in the apparent resurgence of German divisions after Falaise. Thus, curiously, articulation appears to be a form of protection. We should therefore design forces for autonomous action.[52] Articulation as a form of protection is another emergent characteristic of armed forces.

We have seen that combat is not resolved through the outcome of a series of individual fights, but at some earlier point through the individual or collective withdrawal of participation. This discussion of articulation illustrates the fact that motivation and morale are significant factors in combat. We can usefully divide an enemy's motivation or will into its strength, namely 'resolve', and its content, namely 'intent'. A number of strategies are open. Firstly, we can attack his intent: render his aim or mission irrelevant or unachievable. That may be by destroying all or part of his forces, or by seizing and retaining terrain objectives – such as his capital city at the strategic level, or a hill or a bridge at the tactical. We will probably have to fight to achieve those goals. We will probably have to move in order to do so. At the same time, doing those things will tend to reduce his resolve, by promoting the perception that he has failed or is failing.

Secondly, we might attempt to attack his perceptions directly. Deception, psychological operations, operational security, the destruction of command posts and communications systems and several other techniques can be used to deny the enemy his ability to understand the situation. Alternatively they may present him with a false picture. These are generically termed 'Information Operations'. They might aim to persuade the enemy he is beaten. Conversely they may assist the process of beating him, either by denying him the ability to decide until too late, or persuade him to do something that will expose him to defeat.

Thirdly, we can attack the enemy's resolve, his will to continue. We should render him shocked, stunned and demoralized. Heavy concentrated firepower and a high tempo of operations seem appropriate. We will consider this aspect of combat later. For the present discussion we will use 'will' as the determination of an individual to persist with a task despite adversity.[53] If we consider armed forces as human organizations, we can consider the collective intention to achieve or exceed the task to be the sum of the individual wills of commanders at all levels and the cohesion of their troops in following their commanders' direction.

The motivation of the individual to the task is strictly beyond the scope of this book. It is the subject of the Moral Component. We can, nevertheless, say that several factors contribute to it. Physical factors are not necessarily prominent. There are many examples of cold, tired and hungry troops achieving near-miracles, and in some instances fighting to their certain deaths. There are other examples of fresh, fit and well-provisioned forces giving up. We will not investigate that further here. We can say that in the final analysis every soldier makes an individual decision whether to participate or not. That decision may or may not be rational, and may or may not be conscious. However, participation eventually devolves onto personal decisions.

Such decisions are highly coloured by circumstance. Firstly, combat is dangerous. Faced with the threat of death or disfigurement it is entirely rational, although frequently subconscious, for human beings to quit their assigned task in order to protect themselves. It is not surprising that they do; we should study the circumstances in which that happens. Secondly, primates have eyes on only one side of their heads. Humans cannot easily guard against threats from two widely differing directions. In such circumstances it is again entirely rational (if sometimes subconscious) to quit a given task and withdraw to a position whence one can at least observe both threats.

Equally, since man possesses reason, we must expect that in some circumstances he will *believe* himself beaten and quit the task; whether he is objectively beaten (whatever that means) or not. Thus there are several circumstances in which we can reasonably expect real human beings to desist from combat. One might define the tactics of a manoeuvrist approach as being actions which exploit those circumstances. We can also see that certain terrain features might figure in this process. It may be because gaining them allows lethal force to be brought to bear on the opponent. Alternatively it may be because the enemy *believes* that their loss places him at a disadvantage.

Thus it seems quite possible to develop structures and processes that attack the enemy's determination to persist at his assigned task. They may be far more efficient than numerical attrition. More positively, something other than attrition alone normally persuades the enemy to desist. That 'something' appears to be based on attacking the enemy's will and cohesion. Since war is adversarial and the opponent will try to resist his own destruction, an approach that harnesses the enemy's wish to avoid destruction seems more efficient than one which seeks to destroy him, which he will fight to avoid. Thus an approach based on an attack on will and cohesion may be a highly effective way of waging war, and possibly quite efficient.

The key to this is the understanding that armies do not just fight. They do many things, of which fighting is the most important. In general an army should manipulate the use, and manipulate the threat of the use, of violence. It is important for it to be good at both. Since defeat occurs when the enemy believes he is beaten, manipulating the threat of violence is a major tool in winning. It is also important that an army does not 'just fight'; war should be the use of collective armed violence for the purposes of the state. Therefore there should be a clear and obvious purpose for every battle and engagement, which should in some way derive from the aims of the state.

Our discussion of Archer Jones's troop types overlooked a paradox. The model indicates that heavy infantry can normally resist the attack of shock troops. The paradox lies in that, given animal or mechanical effort to assist, shock troops have often been better protected than heavy infantry, not least because they were often drawn from a social elite and could afford better armour. Why, then, have heavy infantry traditionally been able to resist shock troops?

The answer appears to lie in the cohesion that heavy infantry can display. Swiss, Scottish and Flemish spearmen, often with no more than a few breastplates for protection among the front ranks, repeatedly resisted and sometimes defeated armoured knights. Given anti-tank weapons sufficient for the job, there are many instances, from 1940 onwards, of dug-in infantry driving off unsupported tanks, even though a given tank may have been technically capable of destroying a given gun, trench or emplacement. It appears that for (heavy) infantry, cohesion is better protection than armour.

Cohesion appears to have two main aspects: the physical and the social. To Harold's *huscarls*, the need for physical cohesion was obvious: stand alongside your companion, to the death if need be. To a Marlburian infantryman, cohesion implied slightly more. A continuous array of bayonets augmented by volleys of musket shot kept the enemy at bay. Volley fire created holes in *his* cohesion.[54] By 1940 things had moved on. Platoon and company commanders sited LMGs to overlap and interlock.[55] FOOs tied up fire plans to cover the front with indirect fire. Where that was done properly it was remarkably successful. During the Second World War, occurrences in which British infantry positions were effectively broken up, as opposed to being encircled and defeated in detail, were curiously rare.

Further physical aspects of cohesion in defence include the orders for, and rehearsal of, counterattacks. Physical measures generate two things. Firstly, the mechanical aspects of cohesion that are the explicit goal: LMGs *do* interlock and the FOO *can* cover the front with indirect fire. Secondly, the personal confidence born of experience that these things are attended to. Seeing it done on the ground will generate more confidence than classroom lectures.

However, the social aspects of cohesion are arguably as important, if not more so. Soldiers will and do stick together for several reasons that are psychological, institutional and cultural, rather than physical. The Swiss pikemen originally came from closely knit communities given a sense of identity (and social injustice) from their geostrategic position. Flemish spearmen were recruited from city trade guilds.[56] Cromwell's New Model Army had a remarkably coherent religious and political outlook.[57] There are many aspects and many manifestations, from life-long 'buddies' forged on the battlefield, through *esprit de corps* and élan.

They all motivate soldiers and forces to stick together and thereby to persist with the given task. There is also probably a learning, or second-order, effect. Harold's *huscarls* and their forefathers had probably learnt through experience that standing together was tactically vital. It then became a social and cultural value, reflected in expressions such as 'sword brothers' and 'sworn comrades'.[58] The Swiss term for their Confederation is the 'Eidgenossenschaft'; 'Eidgenossen' are those who have sworn an oath.[59] The original oath was sworn at the meadows of Rutli when the Confederacy was first formed of the three forest cantons in 1291. Swiss conscripts continue to swear such an oath on joining the army. The 'Rutlischwur' was, and is, a tangible piece of the identity of the Swiss armed forces.

Some interesting examples reinforce this idea of social cohesion. We cannot be sure that Henry V said the words attributed to him by Shakespeare on the dawn of the Battle of Agincourt:

> We few, we happy few, we band of Brothers
> for he that sheds his blood with me today shall be my brother;
> be he ne'er so vile this day shall gentle his condition . . .[60]

Note that Shakespeare chose to say that the King of England placed himself on a social level with those villeins (the archers) who fought alongside him. He dismounted and fought on foot. Whatever the tactical dynamics of the day, a king who dismounts and fights on foot (and who is ascribed as placing himself socially equal to his foot-soldiers) is making a powerful statement to those infantrymen: we are together. (If anything, Shakespeare was creating an image of a Tudor, rather than a Plantagenet, military leader.)

During the 1980s the British Army produced a recruiting poster entitled 'All the signs of a good career'. It displayed all the cap badges of the army. To an anthropologist, culture is identified primarily through myths, values, symbols and beliefs. The set of symbols that the army chose to display were those of its individual units; not the insignia of the army, nor its directorates (for example, the mailed fist of the Armoured Corps or the bayonet of the Infantry). The army effectively displayed the symbols which in practice discriminate between combat units, and between them and the rest of the army. The poster was obtained by serving officers and NCOs and displayed widely in offices and barrack blocks. It clearly resonated deeply with the army's values. Anthropologically it was a very significant display of what the army felt to be important.

There are many related sets of rituals, myths, symbols and values. In the Israeli army, surviving brothers of a soldier who dies in battle are removed from the battlefield.[61] Although immediately counterproductive in the short term, it is one aspect of what makes the Zahal special *to those who serve in it* and thus contributes to its cohesion. I attended an *Oktoberfest* reunion of former soldiers of the Bundeswehr's 62 Panzer Grenadier Battalion in 1983. The former soldiers were all time-expired conscripts, some of whom had served in the 1950s. All lived relatively locally and some still had a reserve liability. The event was typically

Germanic: oak leaves, pine furniture, beer and 'oompah' bands. It was obvious that, for those current and former soldiers, there was something individual and important about a numbered battalion with no combat experience and no visible identity before the late 1950s.

Turning from the social to the organizational, the German practice of requiring all potential officers to serve in the ranks before commissioning has two useful cohesive aspects. The first is a shared background. Unlike, perhaps, some British units, officers are intimately aware of the dialect, idiolect and attitudes of the soldiers. Secondly, there is shared experience: all ranks have been through many of the same circumstances, and an officer is an officer because he has shown himself to be a better soldier than other recruits of his intake. That is probably a significant benefit.

Cohesion is therefore an issue of interpersonal cooperation, and has mechanical and social aspects. Training and rehearsal can affect both. Training practises the mechanical aspects of a task, which increases the probability that it will go right in battle. As suggested earlier, it supports self-confidence amongst those who train and rehearse. Importantly, it also supports social aspects of cohesion. Troops who train together are more likely to get to know each other, and be able to accommodate each other's strengths and weaknesses, than troops who have never worked together before a battle. Thus at both at an individual and an organizational level training builds cohesion, in both its mechanical and its social aspects. Or, at least, it should. Training, like combat, is a human activity. It can be done badly, and the wrong lessons can be learnt. If so, soldiers may lose confidence in their equipment, their tactics and their leaders.

Similarly, combat experience should build not just skills and collective performance but also social cohesion. Nothing in peacetime really resembles war, not least because of the absence of fear. Within reason, troops who have experienced combat can learn from it, and become both more practised and a more closely knit team. Were that not so, the concept of a veteran unit would be hollow. Yet it does not always happen. High casualties can reduce a unit's effectiveness faster than the positive benefits of experience. And nothing will reduce a unit or formation's ability and confidence faster than being badly beaten, particularly if it also takes high casualties. Conversely, successful operational experience appears to be the strongest source of social cohesion in combat units.

There is at least one good example of where 'veteran' formations were found wanting when next asked to fight. The British 7th Armoured, 51st (Highland) and 50th (Northumbrian) Divisions fought in Africa and Italy with the 8th Army before being withdrawn and committed to Normandy – on D-Day, in the case of the 50th Division.[62] The 7th Armoured and 51st Highland simply did not perform to expectation. Eventually both divisional commanders were removed[63] and a large number of other officers replaced.[64] These measures were largely successful, and both ended the north-west Europe campaign with their proud records restored. The 50th Division did not obviously perform badly, but was one of only two British Divisions broken up in September 1944 to provide infantry replacements.[65] One might expect a 'green' division to suffer this fate in preference; there were seven such in 21st Army Group.[66]

The 51st Highland's return to grace is illuminating. Its new GOC, Major-General Tom Rennie, had commanded 3rd Division but was wounded early in the campaign.[67] Rennie, who had earlier commanded a battalion and a brigade in 51st Highland, cannily appealed to the division's pride in its past achievements[68] (it had fought with distinction in the First World War as well as in North Africa). The twist in the tale is that the 51st Highland Division that fought at Alamein and Wadi Akarit and then Normandy had no direct connection with its First World War predecessor. That became a TA formation which was mobilized in 1939 and captured, largely intact, at St Valery in 1940. Only a few individuals had any connection between the two.[69,70] The new division was largely formed from the 9th (Highland) Division, which was disbanded. It is significant that one Highland formation was broken up to reform another. Rennie's appeal to corporate pride appears to have worked *despite being at least in part false*. In relation to collective military performance, objective facts have only so much relevance. Perception and motivation play a major part.

The poor performance of several formations can be attributed to entirely understandable human and technical factors, once the attendant myths are dismissed. The US 106th Infantry Division was very badly mauled in the Battle of the Bulge in December 1944.[71] It was given the unfortunate nickname of 'the hungry and sick'. There are accounts of soldiers running away, having lost their will and cohesion. Being very newly arrived in theatre, the division was unused to mountainous and hilly terrain, and consequently had significant difficulties making its radios work[72] – a very 'mechanical' aspect of collective performance. The US army became aware of its deficiencies in preparing inexperienced divisions, and revised their training accordingly. The 106th was the last 'green' US division to be mauled in Europe.[73]

That was not trivial; the US army committed 15 further divisions to Europe before VE Day.[74] The problem was, however, largely of the US army's own making. For a number of reasons arising from manpower policy decisions, junior US divisions were used as a source of manpower for those already deployed. When they themselves were deployed, they had not been able to build cohesion within units. For example, the 106th was activated in March 1943, but by the time it deployed in October 1944 it had provided 12,442 replacements for other divisions from an establishment of 14,253.[75] Even worse, the 69th Infantry Division lost 22,235 enlisted men between activation in May 1943 and deployment in November 1944.

It appears on balance that after 1941 or so the German army had mastered the art of generating, sustaining and when necessary regenerating battle-ready formations, whereas (certainly by 1944), the Western Allies had not. To be fair, the Allies had had less opportunity until then. It also appears that much of the Germans' success was due to tangible human issues. Training is probably one. All armies train, but in the German army training and operational effectiveness were apparently given precedence to the detriment of other more pedestrian aspects of military life.[76] This should be so in other armies, but there is continuing evidence that in practice it is not.[77,78]

Local recruiting was another factor. Many armies recruited locally; the Germans seem to have done so more rigorously than some other armies.[79] They also seem to have had a fairly generous leave policy (German soldiers could achieve home leave more readily than their Allied counterparts for most of the war), and a steady trickle of battle-weary officers and soldiers were posted out as instructors. Certainly in the case of the British 7th Armoured Division, there is a view that this should have happened, but did not.[80] Some of those who were relieved of command from 51st Highland in Normandy suffered from combat exhaustion originating in the Desert and Sicilian campaigns.[81]

Cross-posting could have had tangible benefits elsewhere. When the 29th US Infantry Division assaulted Omaha Beach on D-Day, only five of its members had seen combat before.[82] The assault on Omaha was the hardest-fought of the beachhead battles on D-Day, and the 29th took very heavy casualties. Yet 12 US divisions had fought in North Africa and Italy prior to Normandy.[83] The US Army could easily have found a few hundred veteran GIs, sent them to the UK (or even the USA) for leave, and posted them into 29th Division, *if* it had put its mind to it. The 29th was clearly a special case: it and the veteran First Infantry Division were the only US formations to assault the Normandy beaches on D-Day.

Cohesion and collective performance indicate that the whole *is* greater than the sum of the parts. There *is* a systemic effect, and we should see armed forces as systems. Furthermore, most if not all of the components of collective perform-ance reduce to discernible human factors. Those factors include manpower policies, and the rigour with which they are applied; training, and the rigour with which it is pursued; and, in Rennie's case, collective self-image – even where the facts presented are less than entirely true.

All battles have some price in blood. Our revised paradigm of combat requires both armed force and violence, otherwise it would not be dangerous and destruc-tive. Combat is not, however, resolved when all the enemy are incapacitated, but at an earlier point related to the withdrawal of participation. Bringing that about, and bringing it about early, is clearly an efficient approach to achieving tactical success. Thus attacking the enemy's will and the cohesion of his forces, or both, is entirely consistent with many of the ideas we are discussing. It reflects the idea that combat is conducted between human organizations that are themselves complex. It attacks the enemy at a systemic, not an individual, level.

Two extreme and several intermediate cases are possible. In the first, the Commander-in-Chief feels he is beaten and effectively leads his forces into defeat. That might describe Lord Gort in the French campaign of 1940. He withdrew his forces from France in relatively good order and apparently without losing his nerve.[84] Other war leaders did lose their nerve: for instance, the Egyptian Field Marshal Amer, when the Israelis broke though in the Sinai in 1967.[85] Thus one extreme case is where the commander feels he is beaten and his forces respond accordingly.

The other extreme cause occurs where the armed forces give up and withdraw their participation; the high command is left with nothing with which to engage the enemy. This may not be particularly dramatic; the mutinies in the French army in 1917 are an example.[86] There are many intermediate cases; typically where units or formations have withdrawn from participation, partly because their commander has led them away, possibly because he realized he was powerless to stop it. In a messy, unstructured, human activity like war few such instances will be clear-cut.

In each case, we should discriminate between individual will and collective cohesion. Once Gort had decided to withdraw, his army withdrew as ordered. In 1917 entire French divisions remained largely cohesive in their determination not to fight;[87] even though that determination must eventually have been manifested as individual choices by individual soldiers.

Importantly, there is a critical difference between will and cohesion as defined here. In the case of individual will, the rank and authority of the individual is a dominant factor, *particularly* if his forces retain their cohesion. A general can lead his entire army from a battle; a corporal can only lead his section. The commander may order a retreat that turns into a rout. In two cases (the British in 1940 and the Egyptians in 1967) the whole army was effectively removed from battle due to its commander's decision. The Sinai case was the more dramatic, since the Egyptian army also lost is cohesion. When considering the attack on will, we should focus on the enemy *commander's* will; and the higher level, the better.

5

Shock and Surprise

Our paradigm stresses that combat is fundamentally a human phenomenon. In developing that paradigm we looked at tools and models: GST, the Dialectic of Aims and Means, and the Archer Jones model. We discussed the concepts of destruction, protection, participation and cohesion. They all appear useful and informative. However, only 'participation' and 'cohesion' have any obvious human content, and even they are somewhat anodyne. We have not yet seen anything to link the deeply human experience of war with its tangible outcomes of success and victory. That link should be more than just an intellectual sticking-plaster. We seek sound, empirical and convincing connections between human behaviour and battlefield outcome.

Battle is an assault on the enemy's will and cohesion. When seen from a distance, the effect of a number of junior or intermediate commanders losing their will to participate and withdrawing their troops' participation can be viewed as a breakdown in cohesion. The effects of an attack on will and cohesion will not be clear-cut. A systemic view of the problem is required, and a systemic effect sought on the enemy. From a systemic viewpoint, we seek methods that reduce enemy participation in combat to the point at which they feel beaten, surrender or withdraw. Reaching that point quickly generally seems more effective (and also possibly more efficient and economical) than numerical attrition. Yet the destruction of will and cohesion does not just happen. We seek a concrete mechanism which links manoeuvre and weapons effects to a reduction of individual and collective participation. The mechanism appears to involve shock, surprise and suppression.

Although much used by military writers, there was until recently no agreed definition of those terms.[1] British Royal Armoured Corps publications have used 'shock' and 'shock action' extensively, without defining the terms. Artillery doctrine has stressed 'neutralization', which seems to be related. Some British historical analysis (HA) does, however, provide a useful model of shock and surprise.[2] That model is based on hundreds of combat records, from the American Civil War to Vietnam.

HA is a rigorous process of statistical research. It employs professional historians to discover the frequency or incidence of given phenomena from the historical record. Statisticians then analyse that data, using a number of numerical methods. It is a very powerful technique, particularly where the samples are large. Large samples typically mean that findings can be given a clear, and often high, level of confidence. HA can also help unwrap problems where there are many factors at work, and where it is not obvious to the observer (not least the professional historian) what the overall relationships or trends are. HA cannot demonstrate causality. It can identify correlation, leaving causality to be inferred from discussion such as this. If we find HA results from large samples with high confidence limits, we can be reasonably sure that we are observing something important. It is for us to discern just what that is. Figure 12 illustrates the HA model of shock and surprise referred to above.

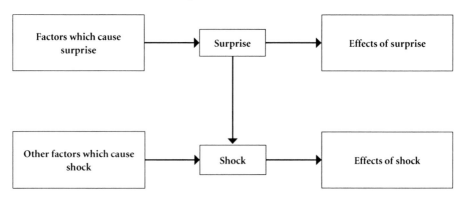

Figure 12: Surprise and Shock

Several factors or influences cause surprise. In turn, surprise is one of several factors that cause shock. Both shock and surprise have effects which can be observed in combat.

Surprise can be generated in a number of ways; most notably through unexpected timing, direction, means or methods of attack. Unexpected timing is especially effective if it is early. Surprise through unexpectedly early arrival can occur at any point, not just at the start of a battle or engagement. Surprise though unexpected direction is particularly effective if the resulting attack is from the enemy's flanks or rear. This may result from the original plan of attack, or from deep penetration and bypassing, which produces unprotected flanks and rears. Unexpected methods of attack typically result from novel tactics, while unexpected means of attack typically result from novel weapons. The Egyptian crossing of the Suez Canal in October 1973 demonstrated surprise in timing, method and means simultaneously. Surprise can be generated in defence through methods such as tactical depth, the concealment and subsequent commitment of reserves, or by the sudden withdrawal to defensive positions in the rear. Deception, intelligence, security, speed and originality are major factors in achieving surprise.

The effect of surprise attacks on the enemy's flanks and rear is considerable. HA of a representative selection of infantry battles showed that in a company- or battalion-level attack, the attacker's casualties tend to be about twice those of the defender if the attack is frontal. Where the attacker manages to find an exposed flank and attack it before the defender can take adequate measures to protect himself, the defender typically takes slightly more than twice as many casualties as the attacker. When an attack strikes the defender's unprotected rear, the attacker tends to inflict almost four times as many casualties as he suffers. The overall advantage of rear over frontal attacks is of the region of seven-fold in terms of casualties. There are very few, if any, other things that an attacking commander can do which will give him an advantage of that size.[3]

Both the probability of surprise occurring and its effects have been quantified. Surprise occurs in about 40 per cent of infantry attacks.[4] It has three main effects. It increases the probability of success, reduces the attacker's casualties, and increases the probability of inflicting shock. The probability of success in the attack in an armoured battle typically ranges from 40 per cent to 54 per cent when there is no surprise. Where surprise occurs, the probability of overall success is about 75 per cent. This suggests that a different mechanism is at work when surprise is present. If surprise is achieved, the probability of success is largely *independent* of force ratio. If success is not achieved, the probability of success is *highly dependent* on force ratio.[5] This indicates quite clearly that if surprise is not achieved, the resulting fight is largely a matter of attrition. The more forces the attacker has, the more likely he is to win. Conversely, if surprise is achieved the defender tends to give up, break or withdraw; and tends to do so much sooner. In infantry attacks which achieved surprise, attackers' casualties were on average 42 per cent lower than where there was no surprise. This figure excludes attacks where shock also occurred; we will discuss these later.

Human factors research suggests a model of surprise and its effects on the behaviour of individuals and organizations in combat. An initial surprise event has four direct and largely physiological effects. They are increased physiological arousal, uncertainty, 'attentional blink' and the cessation of ongoing activity.[6] The surprised person will often fail to remember what they were doing or thinking immediately prior to the surprise event. This is described as an 'attentional blink'. In close combat, these four effects typically last only a few seconds, depending on the extent of the threat. During that time the individual is unlikely to participate in fighting. It might last long enough for an infantry section to lose a firefight and be suppressed, or for a tank to be destroyed.

Over a longer and less immediate timeframe, the effects of surprise are at the perceptual level. Perceptual issues are perceived stress, attempts to reduce uncertainty, and information overload. Attempts to reduce stress by reducing uncertainty are unlikely to be rational. The things which induce most uncertainty are the least likely to be attended to. For example, 'I can't work out what to do about the flanking movement, but I can cope with the frontal threat, so I'll handle that.' The result of these perceptual effects is less effective decision-making.

At the higher organizational level the effects of surprise on a commander and his staff are 'big-picture blindness' and micro-management. History suggests that at higher levels a commander is more prone to big-picture blindness. Due to the immediacy of close combat, at lower levels he is more likely to micro-manage.

Physiological, perceptual and organizational responses can render the individual or the organization more susceptible to being surprised again. The effects of those responses occur over differing timescales. Physiological effects may last only a few seconds and are typically responses to obvious, threatening and immediate stimuli. Perceptual effects occur within a few minutes. They are typically responses to enduring, confusing and distracting stimuli. Organizational effects might not occur for a few hours, and are behavioural responses to unfavourable, disconcerting and undesirable stimuli.

The physiological issues are most applicable to individual combat participation. Perceptual issues are more relevant to low-level commanders; typically at platoon, company or perhaps battalion level. Organizational effects occur at the battlegroup level of command, but are more noticeable in large collective staffs remote from the immediacy of the battlefield, such as at brigade and divisional levels.[7]

Surprise will normally have a greater impact than a force ratio of 10:1. The creation and exploitation of surprise was central to German tactics in the Second World War.[8] It does much to explain the difference in battlefield performance between the German and US armies described by van Creveld.[9] As the Canadian military historian John English put it: 'The German Army was, in fact, an army saturated with surprise. Mobility and manoeuvre were but the respective means to effect it in time and space.'[10]

Thus surprise occurs in only 40 per cent of battles. However, when it does, it results in an increased probability of success and a reduction of casualties, and is more effective than any likely range of force ratios. Surprise is therefore hugely important. Commanders should go to great lengths to achieve it and exploit its results. There are few other things they can do which are as effective in terms of achieving tactical success.

The probability of achieving surprise in a single attack in the battles studied was only about 40 per cent. Importantly, the probability of achieving *some* surprise in a number of separate attacks can be much greater. That observation comes from a simple simulation in which the attacker initiated four simultaneous low-level attacks on separate axes.[11] It demonstrated how the probability of achieving some surprise varies with the probability of surprise in any one attack.

Unsurprisingly, the probability of achieving some surprise (in at least one of the four attacks) is always higher than the probability of achieving surprise in one attack. The simulation indicated that in practice it should not be difficult almost to guarantee the achievement of surprise somewhere. A slightly less obvious finding was that the chance of achieving multiple surprise (that is, surprise in two or more attacks) can be quite high. That is important. By using tactics which seek to achieve surprise on a number of separate routes simultaneously, the attacker should be fairly sure of *some* success; and reasonably sure of *multiple* successes. As we shall see, a number of threats in combination may systemically crack will and cohesion in a manner which is unpredictable, but potentially highly beneficial to the attacker.

Shock has two main battlefield effects. The first is to reduce defence effectiveness, as measured by the attacker's casualties, by about 40 per cent. This is in addition to the effect of surprise. Hence an attack which achieves both shock and surprise will degrade defence performance by about 60–65 per cent overall. The second aspect of shock effect is to *disrupt* the defence. Some individuals and small groups may continue to resist. Others will give up, surrender or withdraw. This effect will be variable and unpredictable. The overall effect will depend both on exploitation by the attacker and the defending commander's ability to restore the cohesion of his forces. Thus any action that is likely to inflict shock, such as a sudden bombardment, should be followed up by rapid exploitation on multiple routes. That exploitation should locate weak points where individuals and small groups have been shocked, and exploit them.

The principal causes of battlefield shock include surprise, rapid bombardment, sudden approach, the use of armour, and the use of certain types of weapons. Surprise attacks by infantry have a relatively high probability – about 50 per cent – of incurring shock on the defence, but only in conjunction with one of two other factors: weak nationality factor, and night or poor visibility. 'Nationality factor' refers to a residual variation in combat effectiveness that cannot be ascribed to any cause other than nationality.[12] With neither of those factors, the probability of incurring shock is only 18 per cent. Therefore, in many circumstances, surprise does not contribute to achieving shock in infantry combat. However, the achievement of surprise at night, or from close range in close country or poor visibility, is reasonably likely to cause shock.

The temporary reduction of defence effectiveness by artillery is normally called 'neutralization', but can be seen as the shocking effect of indirect fire. Another HA study looked at the amount of high explosive required to reduce defence effectiveness immediately after bombardment by 90 per cent for bombardments of differing lengths.[13] The targets were infantry in good defensive positions during the First and Second World Wars. Their effectiveness was assessed in terms of casualties inflicted on the attackers. The key finding was that short, intense bombardments are much more effective (and efficient) than protracted ones. For example, a bombardment of six minutes can be as effective as one of 100 hours, but only use about 3 per cent as much ammunition overall. Very large numbers of guns are required to deliver the shells at that rate. This partly explains why bombardments in the early part of the First World War were generally less effective than later in the same war. The key factor is the sheer speed at which the ammunition can be delivered. Neutralization by artillery fire is an example of the shocking effect of sudden violence.

The use of armour has two effects. There is some shock effect due to speed of approach: the frightening effect of armour rushing towards the defender. Dive bombers have a similar effect. Shock is also likely to occur when armoured units achieve surprise. This is especially true when the defenders think that the tanks are invulnerable; if the attack takes place in poor visibility; or if it is against a defender with a low nationality factor. The probability of tanks achieving shock can reach 95 per cent in some circumstances.

Shock therefore appears to be the result of a combination of factors which effectively results in a group or groups of enemy withdrawing their participation from combat. It can involve the sudden, concentrated application of violence, rapid arrival, close combat and rupture. It may involve night and poor visibility. Shock is much more effective than the attritional effect of the weapon systems involved. It is an individual and organizational response to a successful attack on will and cohesion. In short, it is a human condition.

Each individual ultimately chooses whether to participate or not. That choice may be neither rational nor conscious. We have seen that articulated groupings are less prone to catastrophic collapse, and that decentralization of C2 is correlated with high morale. Articulation may be inversely correlated to susceptibility to shock. Thus there is a complex and unpredictable relationship between social and organizational cohesion, decentralization and effectiveness.

Shock is clearly important, but what is it? A medical analogy is useful. A person can be said to be shocked if he is numb, lifeless or unresponsive, or perhaps behaving irrationally. By analogy we can describe a military group or formation as 'shocked' if its effectiveness is reduced: its soldiers are not participating, but cowering in their trenches, not firing back, or perhaps running away. There may be panic and collapse. They and their commanders have lost their will, their cohesion, or both. Commanders are either not commanding at all, or issuing orders that are irrelevant to real events. These are the sorts of descriptions we read about in accounts of battles. This is the sort of human condition which we wish to inflict. It may be local; it may be transient; and it may not be predictable.

It is, fundamentally, human. We shall call it shock effect. We will discriminate between shock *effect* and shock *action*, which is all of those mechanisms described in the previous paragraphs. Very simply, shock action reduces to the sudden, concentrated application of violence. In the right circumstances, and particularly in connection with surprise, it is hugely significant.

Shock effect only happens to a defending force, but the identity of defender and attacker varies with immediate tactical circumstances. The moment a counterattack is launched, the attacker becomes the defender. This suggests the value of local counterattack. The onset of shock appears to be sudden in most, if not all, cases: it is semantically difficult to identify shock if not by recognizing the sudden onset of reduced participation.

Suppression can be caused both by direct- and indirect-fire weapons. The tactic of 'leaning on the barrage' used in the First and Second World Wars relied on the enemy being at least suppressed, and possibly also shocked, by the artillery until the assaulting infantry arrived. Almost any lethal object impacting or passing near an individual will reduce participation temporarily. This will be exacerbated by the weight, signature and reputation of the source of the fire. The effect is markedly temporary – it passes soon after the fire stops. Thus 'winning the fire fight' is tangible. Lieutenant-Colonel Lionel Wigram remarked in Sicily in 1943 that either British Bren Guns or German MG42s were heard in battle, but never both together.[14] Yet relatively little is known about the accuracy or weight of fire required to achieve suppression.[15] There is some suggestion that accuracy is more effective than weight of small-arms fire in achieving suppression, but most evidence appears anecdotal. Its use in supporting the close assault is recognized, but other applications, such as in supporting infiltration, are not well explored. Suppression would benefit from further study to link it explicitly to psychological and mechanical phenomena.

It not easy, nor necessarily useful, to attempt to separate shock, surprise and suppression in combat. Causes and effects are not easily discernible. The local effect of, say, suppressing a trench may be to reduce the physical cohesion of the defensive position, allowing attackers to penetrate to where they can destroy that trench. That might shock the occupants of another trench and surprise the local enemy commander. He is thus less able to maintain the cohesion of his troops, and so on. Whilst that passage was a causal description, it is unlikely that real events will be so clear-cut; at least when seen from a distance. Even quite recent accounts of combat rarely display any clear-cut, categoric, cause-and-effect descriptions of close combat.

Suppression, surprise and shock are all transient and of themselves innocuous. It might be unpleasant to be surprised, but the only thing that is damaged is one's equanimity. It is the *exploitation* of shock, surprise and suppression which allows the initiator to inflict defeat. Exploitation further reduces the enemy's will to participate, and breaks apart physical and social cohesion.

Suppression, surprise and shock are all transient. They are also localized to some extent, and the location affected is not necessarily predictable. Artillery fire falling across an objective may well suppress the objective whilst it falls. It may not, however, neutralize all the occupants. Some may resume the fight as soon as the fire lifts. Similarly, the objective may not contain all the enemy. Others to the flanks, the rear or even in front may engage the attackers, possibly even before the fire lifts. The net effect will depend on a large number of factors such as the strength of the defended position, distance from the indirect fire and personal bravery.

Thus in exploiting suppression, shock and surprise we seek to generate and sustain an advantage in a situation which is not knowable in advance, which is inherently complex and in which time is of the essence. Given that it may only take a few seconds for an infantryman sheltering in the bottom of his trench to man his weapon and fire, 'timely' means 'extremely quick' in this case. We need tactics and structures that are decentralized and highly flexible, but which can exploit the real situation in accordance with higher intent. Pragmatism will be essential.

It is clearly possible to use movement and weapons in ways which exploit suppression, shock and surprise. For example, the use of armour in built-up areas could be either very expensive[16] or highly beneficial. The careful shepherding forward of tanks by infantry was practised in 21st Army Group in late 1944. The infantry would flush out hand-held anti-tank weapons, allowing the tanks to get forward into fire positions. The tanks could then destroy MG nests and suppress enemy small-arms fire, allowing the infantry to get forward. After a while the process worked very well, and to the defenders appeared undefeatable, which was one of the factors identified as contributing to shock.[17]

It seems appropriate for a commander to consider the enemy's will and cohesion as targets, rather than his vehicles, trenches and soldiers. We have sought an understanding of combat, and how to wage it successfully, that attacks the enemy force systemically. We have also sought to understand the dynamics of the battlefield in terms of observable and comprehensible aspects of human behaviour. The concepts of shock, surprise, suppression, reduced participation, collapse, surrender and flight seem to provide that understanding. We can now attempt a synthesis of those issues, as in Figure 13.

Enemy troop movement and weapons effect results in individuals being stunned (suppressed and neutralized) or demoralized. This is seen as reduced participation, and possibly panic and flight. These symptoms are all more or less observable at the level of individuals and small groups. At the organizational level, the impact of troop movement and weapon effects is seen as shock effect:

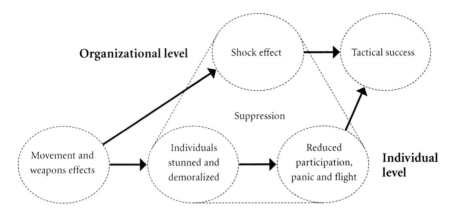

Figure 13: Will, Cohesion, Participation, Surprise and Suppression

the enemy force is collectively numb and lifeless, or perhaps behaving irrationally. It becomes less effective. Reduced participation, panic and flight at the individual level, and shock effect at the organizational level, may lead to local collapse and then tactical success for one side. Suppression lies somewhere in the middle of this scheme. It results directly from weapons effects. By definition, it implies reduced participation. But does it cause loss of will, or does loss of will render the soldier more prone to being suppressed? This model of combat is not entirely clear, and is unlikely to become so, because of the complexity of combat. It is high-level, abstract, fuzzy and imprecise. Although we can identify the terms and place them in a diagram like Figure 13, the relevant phenomena will occur at various times and places on a real battlefield. The real situation will tend to be messy, confused and apparently unstructured. The challenge to the soldier lies in exploiting that situation purposefully.

The attacker should seek to break the enemy's will and cohesion by reducing his participation. Whether the immediate cause of that is shocking, suppressive or surprising to any one individual is relatively unimportant. It is more important to discover a practical combination of actions which systemically breaks the enemy force's will and cohesion. A pragmatic and empirical approach will be needed. We can suggest three useful guidelines. First, surprise is hugely important. Thus whenever engaging an enemy, do so in ways which in practice he finds unexpected. Secondly, when applying violence, do so in a sudden and concentrated manner. That is shock action. Thirdly, because the effects of shock, surprise and suppression are transient and localized, exploit the results. In short, inflict shock and surprise, and exploit the results.

Shock and surprise are everyday, human phenomena. They occur commonly on the battlefield in violent and often lethal ways. Where they do, they are strongly correlated with battlefield success and failure: more so than almost any other condition. Most importantly, they are usable. I taught them at the British Army's Land Warfare Centre from 2002 to 2005, to students taking up appointments as battalion and brigade commanders. The concepts clearly helped students focus their tactical planning to achieve better plans. If doctrine is that which is taught, better doctrine should be that which teaches better things. The concept of shock and surprise appears, pragmatically, to be very useful.

A theory of land combat should do two things. It should give a better explanation of phenomena which we have already observed. It should also allow prediction of phenomena yet to be discovered. It may possibly also even suggest why such phenomena will occur.[18] We should not, however, be too optimistic. We have already suggested that combat may be determined, but it is not useful to consider it so: at least at our present state of knowledge.

Suppression, shock and surprise may be indistinguishable; but ultimately they may all lead to collapse. In the Sinai in 1967, and also in France in 1940, local shock and surprise sometimes led to panic.[19,20] There was flight from the battlefield, sometimes ordered from above and sometimes not. A civil engineer sees collapse as either slow and progressive or sudden, catastrophic and dramatic. It is clearly highly desirable to bring about a catastrophic collapse in an enemy force, if possible. We should attempt to understand how that occurs and how to bring it about. We have seen that mechanisms of surprise and shock action bring about reduced participation in ways which are localized, transient and not predictable in detail. Once we have brought about some local collapse due to broken will and cohesion, the effects of Clausewitz's dialectic come into play. Part of the enemy commander's means is destroyed. The attacker now has all the advantages we identified earlier. If the collapse is exploited, more and more damage can be done. The physical cohesion of the defence is broken. More individuals lose the will to fight. The social bonds of cohesion begin to break down. Panic may become infectious, and collapse becomes widespread. Such a collapse can be highly efficient: after some initial damage, the collapse progresses rapidly for relatively little further input. That input is, broadly, 'exploitation'. This is the kind of efficiency we seek: effective defeat at relatively little cost.

The Dialectic of Aims and Means also indicated the importance of protection. This helps us see the benefit of initiating events. Since both protagonists must protect their own forces, the side which initiates violence is at a systemic advantage. His opponent must *protect* himself and his forces. This will often force him actively to *defend* himself, which tends to reduce his ability actively to inflict damage on the enemy. The initiator will also tend to inflict surprise, rather than suffer it. Surprise is of itself beneficial and may also contribute to inflicting shock.

'The initiative' was defined in British doctrine of the 1990s simply as 'setting or changing the terms of battle by action'.[21] That might be an adequate description, but now appears an understatement. Being able to dictate the course of events is an important prerequisite to winning battles and engagements. Once the initiative has been obtained, it is in principle simple to maintain. If one continues to damage, or threaten to damage, the enemy, he is forced to continue to protect himself and therefore, normally, to act defensively.

This is one important aspect of pre-emption.[22] Pre-emption gains the initiative. It also tends to develop the situation in ways that are broadly recognizable to the initiator, but not to his opponent. The latter will be forced to react to an increasingly unfamiliar situation, and may suffer from the physiological aspects of surprise. Thus he is at a practical disadvantage, because he is forced to defend himself. He also has to react in ways he had not intended. If he has contingency plans and reserves, the effects may not be marked. Conversely, if it is an unexpected contingency – a surprise – the results could be sudden, dramatic and conclusively bad. 'The initiative' might be better defined as 'the ability to dictate the course of tactical events'. It is highly important. Commanders should go to great lengths to seize and retain it, and to recover it if lost.

If one side initiates combat, his opponent must *protect* himself and his forces. But he need not necessarily *defend*. If he can force his attacker to protect himself, he can wrest the initiative back. The pre-emptive attack, or early counterattack, may well seize the initiative from the putative aggressor. Since the original aggressor must also act to protect his forces, he will typically stop attacking in order to ward off the counterattack. Thus the momentum of his attack is reduced or lost. This gains time for the defender to reassess the situation, possibly at a higher level. HA studies stress the relative effectiveness of a defence based on the counterattack in many, but not all, circumstances.[23] It may be locally costly, but overall that may represent a small price to pay. The effectiveness of aggressive low-level counterattacks is a good example of a systemic effect.

This systemic effect may be overlooked. The British historian Paddy Griffith derided the effectiveness of German counterattacks on the Western Front in the later stages of the First World War.[24] Yet even if the British did become adept at beating off German counterattacks by the middle stages of the third Battle of Ypres in 1917, Griffith acknowledges that the British had adopted 'bite and hold' attacks with limited tactical objectives – not least due to the *systemic* effects of German counterattacks.[25]

The counterattack is also one form of operation which the defender can plan and rehearse, so that when he executes it the situation begins to develop in accordance with his mental picture, not the attacker's. Thus once he initiates a counterattack the perceptual advantage lies with him.

'Attack is the best form of defence' was described earlier as paradoxical. With our revised paradigm of combat we now see it to be entirely comprehensible and rational. We begin to see what Thomas Kuhn described as evidence of a better paradigm: one that better describes observed phenomena.

We previously considered that in some circumstances attrition may be an entirely rational and beneficial course of action. In order to be so, it must be markedly one-sided. It should also be as efficient and economical as possible. Armed forces are typically numerous, and if it is necessary to expend millions of rounds of ammunition or gallons of aviation fuel per dead enemy, the war may be too costly to pursue, grievous though the enemy's losses may be. The US involvement in Vietnam may be a case in point.[26]

The greatest criticism of attrition is probably not the somewhat limited criticism inherent in British doctrine (that it may be mutual), but that it tends to be indecisive. Much of Archer Jones's work suggests that shock is necessary for combat to be decisive. Verdun could have bled the French army white, but *of itself* that was probably never going to bring the war to a close. Conversely, it can reasonably be suggested that although the *Kaiserschlacht* was expensive (239,800 German dead and wounded in two weeks,[27] compared with 200,000 in the four months of Verdun)[28], it did at least offer the chance to defeat a British army directly. That might have led to the defeat of the BEF as a whole and an end to the campaign, if not the war.

We have placed a large premium on initiating combat, and suggested the need then continually to damage, or threaten to damage, the enemy. That implies generating and sustaining a high tempo of operations, since combat is adversarial and dynamic. The enemy will wish to seize back the initiative, and will try to do so quickly. To prevent him one must either employ absolutely overwhelming force (which is rarely achieved), or act so quickly that in practice the enemy cannot pre-empt you.

We have seen that there is considerable benefit in thinking and acting faster than the enemy. We can now see other advantages of high-tempo operations. One aspect of surprise is the unexpected, and this includes being unexpectedly early: a consequence of moving fast. This effect is heightened if one reaches unexpected positions unexpectedly fast. However it is caused, surprise contributes to shock. Shock may also be achieved directly through rapid approach. Thus high-tempo operations can also help to achieve shock and surprise.

Manoeuvre is defined as movement relative to the enemy to achieve a position of advantage.[29] Which positions? What advantage? If we generally prefer a strategy of attacking the enemy's will and cohesion through suppression, shock and surprise, we can see several alternatives. They include positions from which suppressive fire can be delivered, or from which shock action can be inflicted, and positions that support the use of poor visibility, such as at night or in close country, in order to achieve shock.

They also include positions that enable us to surprise the enemy. That provides two options. The first is an unexpected position: 'The Ardennes are impenetrable to tanks!' The second is the enemy's flanks and rear. Every position has a front, flanks and rear at some level. The wish to mass fire in a particular direction tends to mean that weapons and sensors are oriented primarily in one direction. Other directions are by definition the flanks and rear. The importance of flanks was well understood by the Germans in the First World War: in 1917 Hindenburg expressly forbade frontal attacks.[30] Even earlier, in 1903, Colonel Balck stipulated that tactical decision should be directed towards the enemy's flanks and rear.[31] In part this reflected German experience in the Franco-Prussian War, contrasting the Prussian Guard's success at Le Bourget with its heavy losses at St Privat.[32] What was remarkable in 1917 was that such decisions should be sought well below regimental level, unthinkable in 1914.[33]

Attacking the flanks and rear creates several advantages. Combat is fractal to some extent. The higher commander's position also has flanks and a rear, and they will typically coincide with those of subordinates. Finding and attacking the flanks and rear tends to unpick the cohesion of a position, enabling the same or other attackers to find the flanks and rear of a higher-level enemy, leading towards collapse at progressively higher levels. This can become a systemic effect, broadly what Liddell Hart described as the 'expanding torrent'.[34]

'A position of advantage' can now be seen to be one from which the enemy can be surprised, suppressed, shocked or assaulted. Alternatively it may be one whose loss leads the enemy to *believe* he is defeated. On the night of 11–12 June 1982, 42 Commando, Royal Marines assaulted Mt Harriet in the Falklands substantially from the enemy's rear. The battle was over in about five hours and most of the fighting took less than two and a half hours.[35] This contrasts with the other five battles in the Falklands conflict which all took over seven hours' fighting. The unit suffered only two dead and seven wounded. It seized the summit of Harriet early in the battle and thus, physically and psychologically, placed itself above most of the defenders and cut off their retreat (see Figure 14). Over 300 prisoners were taken by 42 Commando, more than any other unit involved (except for the Second Battalion, the Parachute Regiment, who captured 1,007 Argentinians *after* Goose Green when the garrison of Darwin surrendered).[36] It seems that on Harriet the Marines used surprise to seize a position that the enemy *believed* gave them a material advantage, and that following this the enemy rapidly surrendered.

In general, ground manoeuvre and close combat provide the means for the physical possession of the battlefield which is required to resolve the battle or engagement. Manoeuvre should be used to obtain positions of advantage from which one can deliver suppression, inflict surprise and shock, and hence shatter

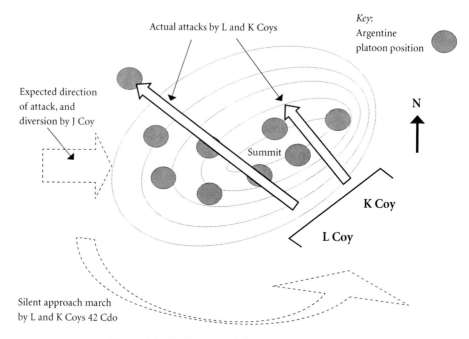

Figure 14: 42 Commando's Attack on Mt Harriet

the enemy's participation. That should bring about a local collapse which should be exploited, not least through further manoeuvre, in an accelerating process.

Combat is dangerous and adversarial. The danger stems principally from the enemy's fire, and that fire generally intensifies as one approaches the enemy. Thus one will often need to beat down the enemy's fire in order to get to that position of advantage, and that is the role of fire support.

Fire may destroy, neutralize or suppress. Destruction typically takes a lot of ammunition, not least because a sensible enemy protects himself from the effects of fire. Neutralization and suppression are transient. Hence, in the general absence of overwhelming superiority, manoeuvre will either require luck (to find an undefended route), surprise (to find a route that the enemy is not covering at that moment), or fire support, in order to reach a position of advantage. Thus fire support is a deeply integral part of combat. We can make several observations.

Firstly, given how lethal modern weapons are to troops moving in the open, fire support should effectively suppress all weapons that bear on the attacker. Given the range of such weapons, a large number can cover a given point, so a large number may need to be suppressed. Alternatively one can choose covered routes, which may conceal short-range enemy weapons. Suppressing these calls for very precise control and rapid response from suppressive weapons. That in turn calls for very low-level decentralization of at least some suppressive fire elements. Put differently, every low-level commander in combat should have access to precise and flexible fire support; unless, perhaps, his army always relies on massive, closely programmed barrages.

There has been considerable discussion of British attack tactics in the Second World War.[37] Perhaps the overwhelming lesson is that against a competent enemy massive fire plans could be largely successful, particularly if the attackers 'leant on the barrage'. Regrettably, 'largely successful' was not enough. The Eighth Army reported many cases of brigade attacks, typically with four infantry companies in the first echelon and supported by eight field artillery battalions and four of medium artillery, which failed to carry the objective because of a few surviving machine guns and mortars.[38] Assuming that at least some of the lessons from the Second World War are still valid, every low-level commander should have access to precise and flexible fire support.

Suppressive fire may have other unintended, unpredictable and beneficial effects. It may, for example, suppress an enemy commander who is trying to organize a counterattack. Some effects may also be detrimental. Armies go to great lengths to avoid casualties due to friendly fire. It is questionable whether, given the intensely complex nature of combat, such casualties can be, or even should be, eradicated entirely. To the German infantry in the latter part of the First World War, a few friendly casualties were preferable to being pinned down by the enemy due to an inflexible and overcentralized plan.[39] Should intense combat ever be encountered on any large scale again, such a costly lesson might be rapidly relearnt.

Exploitation is another tactical term which is not well described in doctrine. Exploitation is largely the seizure of opportunity. It may be planned or opportunistic. Opportunistic exploitation is the seizure of an unforeseen chance; it typically requires action beyond the given mission. For example, a commander ordered to neutralize an enemy force covering the approaches to his commander's objective may find an approach to that objective which is not covered, and simply seize the objective himself. Since combat is complex, dynamic and unpredictable, such opportunities might occur anywhere and at any time. Since seizing them may give him the initiative, a commander should constantly search for such opportunities and pursue them ruthlessly. Exploitation should be expected from subordinates. They should not have to be told to exploit; and only told how far to exploit if absolutely necessary.

At the tactical level, exploitation has concrete, physical aspects. It usually requires the use of manoeuvre, fire, or both. Opportunistic exploitation allows unforeseen tactical advantages to be turned into operational success. It requires commanders with initiative, determination and a readiness to do the unexpected.

Aggressive exploitation is a key mechanism for translating local success into campaign victory. Once the enemy is found, force is applied to fix and then strike him, creating opportunities for higher levels of command. Reconnaissance forces, echelon forces, reserves, or a combination of them, should then exploit rapidly. Reconnaissance should be extensive, expansive and continuous in order to find the opportunities for exploitation. Aggressive exploitation can be conducted at any level at which seeking the enemy's positions, his strengths and weaknesses can be coupled fluidly to the ability to apply sudden concentrated violence. That might even be the platoon or the section level.

The effects of shock are likely to be local, temporary and unpredictable. Localized shock effect should be expanded through exploitation to encourage collapse and paralysis at higher levels and over wider areas. Planned exploitation is designed beforehand to follow anticipated success, and may require fresh, echeloned forces. Opportunistic exploitation is a key mechanism for seizing and retaining the initiative, and building cumulative success at successively higher levels. Opportunistic exploitation should be carried out with all available forces to hand and initiated as soon as an opportunity is recognized, particularly at low tactical levels.

We can now begin to understand the relevance of some of the basic tenets of published doctrine. For example, consider the Dialectic of Aims and Means. Since war primarily concerns eliciting and satisfying aims derived from the objectives of the state, the first principle of war should be the selection and maintenance of the aim. If not, the operation or campaign in question is irrelevant to the purpose of the war.

We can consider the nature and the application of the means separately. A commander's primary concern with regard to the *nature* of the means should be to generate and sustain a fit instrument. He should maintain the morale of the troops, and minister to their physical needs; both in terms of logistic sustenance and that they be administered effectively. He should also protect (secure) them against the enemy's actions.

The *application* of the means in an adversarial, dynamic and dangerous environment requires the commander to concentrate the effect of his force; be economical with forces away from that concentration; achieve synergy through cooperation and surprise where possible; be flexible in application; and strike aggressively.

We can therefore define the Principles of War from the Dialectic of Aims and Means. They are: the Selection and Maintenance of the Aim; the Maintenance of

Morale; Administration; Security; Concentration of Force; Economy of Effort; Cooperation; Surprise; Flexibility; and Offensive Action. These are not absolute. They are an interpretation of the requirements of combat in an English-speaking paradigm. Unsurprisingly, the American Principles are similar.[40] One would expect a different Russian cultural interpretation, yet the Soviet 'Principles of Operational Art' can also be deduced from the Dialectic of Aims and Means.[41] In general, principles (even the much-vaunted Principles of War) should follow from an understanding of the nature of war, and not replace it.

We saw that the Core Functions (Find, Fix, Strike) and the Functions in Combat (Command, Manoeuvre, Firepower, Protection, Information and Intelligence, Combat Service Support) were stated authoritatively in British doctrine, with no justification and no link between them. We can now see those Core Functions to be part of a description of combat which considers war to be adversarial and dynamic. The enemy will conceal himself, to resist the effects of our weapons and conceal his plans, so he must be *found*. He will resist destruction, and attempt to damage or defeat us, so must be *fixed*. Finally he must be *struck* to inflict the damage that both reduces his ability to damage us, and grants us freedom to achieve our aims. The Core Functions do not replace an understanding of conflict, but they can be a useful shorthand or high-level guide.

Fixing the enemy has at least two components, psychological and physical. The physical seems simple: if an enemy is under fire he will have difficulty in moving, as he would if he has to cross a river with no bridges. Fixing by firepower or terrain is fairly obvious and mechanical. Being fixed psychologically is less easy to describe. It can be seen as inducing the enemy to persist with something he is predisposed to do.[42] There is a further case that spans the physical and the psychological. When we discussed Clausewitz's dialectic we stressed the importance of protection. It follows that threatening or attacking something the enemy feels he must defend is a significant element of fixing. Verdun is a prime example: Falkenhayn rightly believed that the French would feel obliged to defend it to the last.[43] To that extent they were psychologically fixed, using a very physical method. Often the easiest way to fix an enemy is to threaten to attack something he has to protect, such as his forces. Deception can play a major role. Deception may fix him until the deception is exposed, which may be too late for him to regain the initiative.

An experienced wargamer once observed that in his experience the enemy almost never enters 'killing areas' designated by players who are serving or retired army officers.[44] Of the five or six such areas designated in three weeks of exercise at battlegroup, brigade and divisional level on Exercise IRON HAWK in September–October 2000, the (live) enemy entered such a killing area only once.[45] It appears that an element of compulsion is missing from British understanding of war. Unless the enemy is predisposed to enter a killing area, he must be forced into it. That will typically require the application of violence. If not, he is not fixed. If he is not fixed, a tactic relying on 'killing areas' will not work. If he is fixed, such a tactic may be unnecessary.

To strike is to inflict the damage which allows the commander to achieve his aims. It will typically be some combination of manoeuvre and violence. The guidance we have developed for shock and surprise applies here: do things which the enemy does not expect and, when applying violence, do so in a sudden, concentrated manner. In some circumstances it may not be necessary to apply violence. A bold manoeuvre may be sufficient. However, combat is a violent business; the notion of a bloodless victory is often a chimera. We have already discussed the relevance of exploitation. It was added to British doctrine as a fourth Core Function, largely for the reasons given above, in 2005.

The Functions in Combat (Firepower, Manoeuvre, Information and Intelligence, Command, Protection and Combat Service Support) begin to break out the Core Functions in terms of tasks for given troop types. For example, Firepower can be used to fix, and is often delivered by artillery units. So we can begin to make sense of some of the slices of doctrine which we criticized earlier as being incoherent.

Archer Jones's model of troop types can be extended to envisage zones of operations for different troop types, cooperating in time and space to achieve the overall intent. Jones suggested a need for shock action in order for combat to be decisive. In the first instance we see that as a need for close combat. Our discussion of fire support suggested that close combat should be a complex interaction of manoeuvre and fire support, from the very lowest level upwards.

With the range of modern weapons, including aerially delivered weapons with ranges of several hundred kilometres, we can affect the enemy far beyond the range of close combat. Traditionally there have been two roles for such distant operations. We can describe them as long-range fires and long-range raids. Such long-range operations harass, delay and interdict the movement of enemy forces to the immediate battlefield. They also suppress, neutralize and possibly destroy enemy indirect-fire systems and air forces. Examples of the latter include the Israeli air force's pre-emptive strikes against Arab air forces in the Six-Day War in 1967, or the use of Tornado bombers to destroy Iraqi aircraft on the ground in the early stages of the first Gulf War. The original role of the Special Air Service in attacking Luftwaffe airfields in North Africa would be a good example of long-range raids.[46]

Archer Jones's analysis would see this capability as opening up an extensive zone of 'missile', as opposed to 'shock' action. Long-range fires would meet his description of missile troops; raids as the actions of raiding troops. The latter have also played an important role in scouting. This would be seen to persist, be it through reconnaissance aircraft, long-range reconnaissance patrols (the original role of the Long Range Desert Group)[47] or, today, reconnaissance satellites.

Modern technology has opened up new possibilities. From the earliest stages of the Second World War, armies delivered significant forces by parachute and glider. Given their limited numbers and mobility once landed, that was generally in order to seize objectives prior to an advancing force. The scale of such an operation might be huge: Operation Market Garden (the Arnhem campaign) attempted such an operation at the level of an army group or even the entire theatre of operations.

Helicopters have extended that possibility yet further since 1945. The US army was experimenting with air mobility up to divisional level in the 1950s.[48] Yet, despite helicopters' mobility, their potential is still limited. Even in the first Gulf War the role of the US XVIII Airborne Corps, which took part in Market Garden 45 years before, was essentially one of raiding, screening and securing the open flank: historically a prime task for light cavalry. In 2003 the US 101st Air Assault Division was unable to create any opportunity to conduct a large-scale air assault in Iraq because the environment was considered too risky.[49]

The British doctrine of the 1990s described all such long-range fires and raids under the heading of 'deep operations'.[50] It described deep operations as taking place over extended ranges and timescales, and being intended primarily to shape the battlefield for decisive close operations. That is all now entirely comprehensible, in terms of our paradigm of combat expanded through the Archer Jones model.

Subsequent publications suggested a desire to conduct deep manoeuvre;[51] that is, decisive operations at some distance from the primary close-combat elements. This is one aspect of a desire to shift the emphasis from close- to long-range operations in order to strike directly at the enemy's strategic or operational centre of gravity. With this shift the role of close operations is reversed. They shape the battlefield for decisive deep operations. This desire is prompted *inter alia* by the desire to avoid casualties through attritional close combat.

There are logical inconsistencies in this approach which centre on the geo-spatial definition of 'deep'. The issue ultimately relies on one question: can missile fire of itself inflict shock, and hence be decisive? The long thrust of history reviewed by Archer Jones suggests not, but that appears to neglect technological development. It might be that up to now long-range missile systems were insufficiently effective, but have now become so. Certainly, airpower exponents and some artillery protagonists would have one think so.

That is, however, to be doubted; not merely out of scepticism, but by considering our paradigm of combat. It is adversarial, developmental and dynamic, and force protection is a key consideration. Every measure that increases the threat to one's own forces drives the search for a counter-measure.[52] Those counters may be technical, organizational, procedural or simply imitative. Long-range fires or raids directed at the enemy's centre of gravity will result in counters such as protection, concealment, redundancy or perhaps just dispersion.[53] Such long-range fires may well cause attrition; but as we have seen, attrition is of itself rarely decisive. It is entirely sensible to try to attack the enemy's strategic or operational centre of gravity directly. However, if that is to be done by violent means, long-range fire alone will probably not do it. There needs to be some coercion, fixing and shock.

'Deep' and 'close' operations were reasonably well described in doctrine in the 1990s.[54] In my opinion, the terms were well understood throughout the British Army. What was far less clear, however, was any understanding of what kind of operations *should* be conducted at which level of command. There appeared to be an unspoken assumption that every level, at least from battalion or battlegroup level upwards, should indulge in both deep and close operations. The assumption or wish that formations should conduct deep operations reflects wishful thinking to some extent. They often did not have the means to do so. It also reflects an unthinking, fallacious assumption. The army told itself about close, deep and rear operations and then just assumed that something as seemingly important as a brigade or division should indulge in all of them.

In practice the biggest shortcoming of the 1994 description of deep, close and rear operations was that they explicitly linked purpose with place. For example, whilst deep operations took place over large distances, they were primarily intended to find and fix the enemy.[55] By assumption, close operations were intended to strike decisively. This issue was resolved in the 2005 rewrite of high-level doctrine by reserving the terms 'deep', 'close' and 'rear' purely to describe location. Operations should instead be described in terms of their purpose: that is, as 'decisive', 'shaping' or 'sustaining'. The decisive operation is that which should, if successful, lead inevitably to the achievement of the mission. Decisive operations are considered typically, but not necessarily, to involve close combat. This construct of decisive shaping and sustaining operations thus implicitly gave a place to shock action in the battle or campaign.

War is dynamic as well as adversarial and dangerous. This suggests another aspect of deep operations: delay, and hence temporal sequencing. There seems to be little convincing evidence of the destructive effect of airpower over the Normandy battlefield. In most cases the number of vehicles destroyed, and particularly the number of armoured vehicles, was modest in relation to the effort expended.[56] Airpower does, nevertheless, appear to have considerably delayed the arrival of German armoured divisions.[57,58] This effect was significant, given the Allied race to build up the invasion force in Normandy faster than the Germans could oppose it. Only one of the dozen or so German armoured divisions available in the West in June 1944 made contact with the invasion force on D-Day.[59] It took six weeks to get all of them in contact, and air power played a significant role in that.

Rear operations are defined as those required to sustain and protect the force. They typically involve combat service support and some element of rear security, including air defence. Sustaining the force translates directly into maintaining the nature and effectiveness of the means, in terms of Clausewitz's dialectic. Force protection can from that perspective be seen as another aspect of the same thing: maintaining the nature and the effectiveness of the means.

At the start of this book we considered several paradoxes. We saw that the prevalence of paradox indicated that the prevailing paradigm of combat is flawed. If we now have developed a better paradigm, those phenomena should no longer appear paradoxical.

'Attack is the best form of defence' is perhaps the best example. We now understand that seizing the initiative and threatening or actually inflicting damage makes the enemy consider protecting his forces. That may be done by defending, which limits his ability to damage you. Thus, if one's overall goal is to protect one's own forces, then attacking (forcing the opposition onto the defence), may be the best way to do it.

Furthermore, in 1987 OA demonstrated that the defender is at a systematic *disadvantage* in close country (be it woods or built-up areas).[60] It seems that, amongst other things, in close country the defender is generally unable to mass the fire of his weapons, due to very short ranges available in relation to unit frontages. Given their relative protection, if only from view, the attackers can mass forces more safely than is normal. They can therefore isolate and attack small bodies of enemy relatively easily.[61] The overall effect was described as 'counterintuitive'.[62] We would now say that it was contrary to the prevailing paradigm of combat, and particularly that of Fighting in Built-up Areas (FIBUA). In FIBUA the attacker is expected to suffer high casualties. By assumption, the defender will suffer fewer casualties. Conversely it seems that such expectations, formed from experiences of high casualties in FIBUA, are based on ignorance of relative casualty rates. Attacking infantry generally have an advantage of 3.57:1 in terms of attackers' to defenders' casualties in FIBUA.[63]

This subject presents some interesting evidence about paradigms: not so much about what men believe as what they *choose* to believe. That OA was reported to the army, and published in the *Army Training News*.[64] It was factual, numerical and high-quality evidence based on historical analysis and extensive field trials. The army's FIBUA manual was amended two years later, but showed virtually no sign of accommodating that knowledge.[65] This itself was commented on in a further BAR article, which incorporated further evidence from a battalion which had studied and practised FIBUA in Berlin almost exclusively for two years.[66] On that occasion the article attracted absolutely no response from the infantry or doctrine communities. Either the subject did not seem important (which seems unlikely, since the army had only just built a FIBUA complex at Copehill Down on Salisbury Plain at great expense), or the evidence was so far beyond its expectations that it was just not taken seriously.

'Attack is the best form of defence' has two more psychological aspects. Firstly, attacking places the opposition mentally on the defensive. Secondly, as discussed above, the attacker sees a situation unfolding in a way that is broadly in accordance with his view of the situation, whereas his opponent's view of the situation is changing in a way he might not have foreseen. Thus for several reasons, all of which emerge from our revised paradigm of combat, 'attack is the best form of defence' is not paradoxical. It is entirely comprehensible, and consistent with what we now understand of combat.

'If you wish peace, prepare for war' is another apparent paradox. Clausewitz's dialectic suggests that possessing suitable means, and prior consideration of aims, reacts on a potential opponent's aims and means. That suggests some merit in both preparedness and deterrence. Thus 'if you wish peace, prepare for war' is not paradoxical.

General Orde Wingate's motto for the Chindits was that 'the boldest measures are often the safest'.[67] That is seemingly paradoxical, if somehow resonant with what one knows of the Chindits. Yet we can equate boldness with a predisposition for offensive action coupled with surprise. Given our new understanding of why 'attack is the best form of defence' and our discussion of surprise, there is little paradox in it. Indeed, the three apparent paradoxes presented in this section should not now seem paradoxical at all. They all suggest that we have developed a better paradigm; one that resolves paradox and better describes observed phenomena.

An apparent drive towards asymmetry is considered to be a characteristic of modern war.[68] We should be dubious. Much of the discussion of asymmetry is a fad.[69] Firstly, war and combat have always been asymmetric to some extent. No two armies have ever been entirely identical. Even if the means were largely similar it is unlikely that their aims were. Secondly, war is adversarial. The initiator should seek to exploit his advantages, which typically include the *differences* between his forces and his enemy's. The largest single class of apparent symmetry is civil war, where both sides may have access to, or raise similar forces and may have symmetric goals. That is, however, a fairly minor class of exception and does not relate to warfare in general.

Seeking asymmetry is entirely rational, and some degree of asymmetry should be seen as fundamental to war. The real question is 'how much?' Another issue is that war is adversarial and developmental, and it is entirely sensible to seek to counter one's enemy's advantages. That might be by imitating means and methods, which is a tendency towards symmetry. This could be considered as part of a dialectic: on the one side a drive towards symmetry; on the other a drive towards asymmetry.

Which drive prevails will depend on the circumstances, not least the relative availability and affordability of technology. It can reasonably be suggested that the non-availability of high-technology forced the North Vietnamese to consider highly asymmetric means and methods, which they would claim were ultimately successful. In Archer Jones's terms, they used force types which the US army could not counter. Their army was essentially a very large supply of light infantry operating in close country, where the US 'heavy infantry' and 'shock troops' could only operate with difficulty. Jones's model suggests that such battles would rarely be decisive. Yet, as the Vietnamese showed, persistence over more than a generation allowed them to win the war.[70]

If we cannot know in advance whether asymmetric or symmetric means and methods will prevail, we now have tools with which to seek the solution: empiricism and pragmatism. Doctrine should be our current best guess as to how to wage the *first* battle of the next war. We must then be pragmatic and empirical. What works? Find out! Then do it in the second and subsequent battles. And make sure it is the *second* battle; we need doctrinal change which is 'about right, but very quick'.

Our conceptual understanding of combat should allow us better to explain phenomena which we have already observed. It should also allow us to predict phenomena yet to be discovered, and possibly also suggest why such phenomena will occur. That may be optimistic, although in looking at asymmetry we may be able to predict how armies and nations will adapt to technical, organizational and procedural challenges in the future. War is evolutionary.

It would be surprising if anything in this discussion were not largely recognizable from what we already know. However, what has been developed is some concept of 'how' and 'why', which we observed to be absent. For example, we saw that the Core Functions and Functions in Combat were stated authoritatively in British doctrine, but without source or linkage. We have now seen some derivation and some connection. With such an understanding, the relative importance of different elements can be expected to change. The need for pre-emption, surprise and speed now stand out as being key requirements for success in combat. Our improved theory might also predict new phenomena, some of which should be useful.

We have seen several pointers towards the necessary shape, and methods, of ground combat forces. The discussion points towards forces which can conduct aggressive ground reconnaissance and exploit the opportunities that creates. They should be organized and trained around tactics of shock and surprise. They should be able to apply fire flexibly and transiently in order to enable penetration, manoeuvre and irruption. They should be very flexible and responsive: that is, they should be able to decide and act very quickly. They should be especially responsive to low-level opportunities, require only a minimum of orders and hold internal reserves of combat power.[71,72]

We have tried to develop a conceptual understanding of combat. That under-standing is high-level, abstract and imprecise. But in a way which is hopefully comprehensible and credible, it appears to link soldiers' actions with tactical success. It roots warfare in human behaviour, hence in sociology, anthropology and psychology. It describes combat in terms of human beings and the way they act and react, rather than in terms of abstract concepts with no external reference.

Little in this chapter should appear surprising. If our conceptual understanding of combat is valid it may look and feel unfamiliar and different, but it should look and feel right. It should be like viewing the world through a better pair of spectacles. The world looks the same, and as we think it should. However, it can now be seen more clearly, and occasionally we are surprised by seeing things we previously couldn't, or didn't, see.

6

Tactics and Organizations

We now turn to concrete issues of tactics and organizations. We seek an approach to tactics that is empirical, systemic and historically based. How should we best destroy the enemy's will and cohesion? We should clearly focus on shock, surprise and suppression.

Shock should be a mechanism for inflicting massive, lasting defeat. But how to inflict shock is a significant question. We know that shock requires sudden, concentrated violence, rapid arrival and sudden destruction. The tank plays a large part in this: as Guderian put it, '[t]he most valued characteristic of the tank is its capacity to deliver effective and close-range fire against clearly-identified targets, and destroy them with just a few rounds'.[1] Yet this is much more than just a matter of having tanks charge up and drive onto the enemy: adequate for 1940, perhaps, but not even for 1944. Similarly one should seek surprise through unexpected timing, direction and method. We should employ aggressive ground reconnaissance and rapid exploitation. We know that a balance of force types is needed, yet it is not obvious how to create that balance, nor use those forces once assembled.

The dominant theme in discussion of tactics in the 1920s was arguably not that of the tank but rather that of infiltration.[2] Discussion of tanks gained centre stage only towards the end of the 1920s and into the 1930s.[3] The US army's official history of the Second World War, discussing the training of ground combat units, referred to 'infiltration' three times in two pages.[4] The 1942 British infantry tactics pamphlet considered infiltration far more than anything published in the last few decades.[5]

The gist of infiltration tactics is to find and attack weak spots, typically the enemy's flanks and rear, on the immediate battlefield. Suppressive firepower is used to fix, and mobility to turn flanks. Raiding and depth fire are used to isolate the immediate enemy, but not within the contact battle. This process is applied successively at higher and higher levels, but with a difference in emphasis. Combat is not fractal: the way that sections and platoons are handled differs from that of battlegroups and brigades. Nonetheless, the overall effects should be coherent. Section- and platoon-level successes are company- and battalion-level opportunities. Tactical success sets the conditions for operational-level engagements.

Infiltration should be multiple. Attacking on multiple axes should be more likely to deliver surprise. Queuing theory suggests that a single, successive, one-after-the-other approach (for example, with three companies attacking through each other along the same axis in turn) is unstable. The process is likely to break down, and such attacks lose momentum.[6] The attack of the Second Battalion, the Scots Guards on Tumbledown in the Falklands conflict is an example. That of the Third Battalion, the Parachute Regiment on Mt Longdon is another. Both were inherently successive rather than simultaneous. Both took much longer, and both battalions suffered more casualties, than the two Commando attacks on Mt Harriet and Two Sisters.

Infiltration should also be covert where possible, if surprise is to be gained. On some occasions surprise will be impossible, but attackers should revert to stealth as quickly as possible to harness the systemic and emergent advantages of surprise. Similarly, within attacking echelons commanders at all levels should have immediate access to flexible and accurate suppressive fire, to enable manoeuvre.

Both Hindenburg and Patton forbade attacks that were frontal at the point of application.[7,8] 'The point of application' is important. It may at times be necessary to attack an enemy frontally at unit or formation level. But the scheme of manoeuvre should allow for sections and platoons to break into the enemy's positions, find his *local* flanks and rear, and exploit them. If there are no gaps, create them: shock and stun him with artillery. The result will be patchy. Some enemy will be destroyed, some cohesion will be lost and the effect will not be predictable in advance. That will create local and transient opportunities for exploitation.

Exploitation means reinforcing success. Commanders should watch which subordinate is best making progress, and support him with fire. That fire may not be to the subordinate's front. It may be to screen a flank, or break up a local counterattack. Reinforcing success also means pushing reserves forward where success is obtained, even though it may not be part of the intended plan.

The converse of reinforcing success is closing down failure. If a subordinate finds the enemy resisting uniformly across his front he should not persist, but rather pin the enemy with fire and anticipate that he will be re-tasked. His superior will wish to use him to reinforce success elsewhere. At the very least his reserve or his fire support may be taken from him to support success elsewhere.

The key to successful infiltration may well be simplicity. The aim is to confuse and distract enemy commanders with a complex, dynamic situation which they find hard to understand; still harder to react appropriately to. The *enemy*, that is, not one's own forces! From the enemy's perspective, the complexity emerges from multiple attacks, simultaneity and dynamism. It is not necessary to inject complexity into our own plan. It is undesirable to do so. One should use plans that are essentially simple, simple low-level drills and simple contingencies, not least because the situation will become complex. A German military saying has it that 'in war, only that which is simple will succeed'.

Night attacks are a good way of conducting infiltration. In one example of night infiltration from the Eastern Front in 1944, 36 Germans seized a Soviet position held by 100 men with six HMGs, heavy mortars and an anti-tank gun. The attackers sustained one casualty.[9] At least some of the success of the six British battalion attacks in the Falklands was due to the exploitation of darkness. Modern surveillance and night vision devices may have eroded that advantage, but they have also given the attacker some benefits. The Argentinians had modern night vision devices and surveillance radars in 1982.[10] Infiltration and stealthy night attacks were successful in several places in Normandy.[11] On at least one occasion, the assault was the attacking unit's first engagement in the war.[12] Night infiltration is not the sole preserve of experienced units.

Trials of infiltration tactics conducted under controlled conditions showed some interesting results. In one trial in Cyprus in 1996, using a novel structure but a predominantly infiltrationist approach, the attackers took 26 per cent fewer casualties, found the enemy's flanks 50 per cent more often and his rear 400 per cent more often than the same troops using conventional tactics. The average rate of advance was 77 per cent faster. Critically, the rate of approach of the *fastest* penetration was about 16 times faster.[13] The speed of the fastest approach, not the mean speed, has the greatest psychological impact on the enemy commander. Unfortunately, psychological effects could not be measured during the trial.

In another trial in 1999, 40 Commando, Royal Marines undertook a series of similar engagements using a variety of different structures but with broadly similar results. Interestingly, the results of both trials were ignored. The reason appears to be that both the British Infantry and Royal Marine Commandos see infiltration as outside their paradigm of combat. Ironically, the former commander of an Israeli parachute brigade once told me that 'that's the only way to do it'.[14]

Opportunistic deep penetration on multiple axes supported by stunning indirect fire is an ingredient, but cannot be the whole answer. It should be supported by aggressive ground reconnaissance to find initial, higher-level opportunities for surprise and pre-emption. That aggressive reconnaissance almost undoubtedly requires tanks, not to fight for information but to fight! Deep penetration should inflict local shock and surprise, which should be exploited to inflict local collapse. This may achieve local success, which should be exploited further. It will not win campaigns, however, unless the enemy can be prevented from containing local successes. That is the role of 'raiding troops'. The Archer Jones model provides the intellectual framework for deep raids and deep fires to harass, disrupt and delay the arrival of enemy reserves. They can also distract the enemy through deception or surprise.

We can now identify the main role of armour and infantry battalions. They are the instruments of shock action. They apply, or control the application of, numbing firepower in close combat. They should be prepared to drive a long way to avoid fighting where possible, but equally be prepared to switch rapidly to the sudden, concentrated application of violence, exploiting surprise and multiple infiltration. The primary requirement is for enough echelons to support infiltration attacks in depth: platoons to support sections, companies to support platoons and battalions to support companies. They should all have their own suppressive fire capability. They are primarily close-combat groupings. Other units should conduct the deep operations required at higher level.

This discussion may understate the role of the tank. Common practice tends to stress the tank's ability to destroy other tanks, which is logically self-defeating. Clearly enemy tanks need to be destroyed. However, we should remember that the tank was originally intended to help the infantry advance. Both tactics and technology have developed enormously since 1916. A modern 120 mm high explosive or high explosive Squash Head (HESH) tank gun round is devastatingly effective in a way which the soldiers of 1916 could only dream of. Tanks give the armoured portion of an all-arms battlegroup the ability to inflict shocking, near-instantaneous violence on targets which the infantry has real difficulty in destroying, such as machine gun nests and buildings. The real contribution of the tank, like that of the infantry, is the part it plays in all-arms battlegroups which exploit the characteristics of both arms in a synergistic whole.

Thus far we have only considered the attack. Analogy and history suggest a format for the defence. The defender should seek to harass, disrupt and delay the enemy's attack. He should aim to break the attacker's will and cohesion, and seize back the initiative. This will not be done by waiting passively for his attack, nor by carefully coordinated fire alone.

The defender should counterattack at every level, and as early as possible. Counterattacking breaks the attacker's momentum and disrupts his understanding of the situation. It breaks his psychological advantage. Counterattacks depend in large part on surprise for their success.[15] Early low-level counterattacks should find local flanks and covered approaches, and achieve surprise. 'Immediate' counterattacks organized at high level tend to arrive too late, at least in woods and forests.[16] This is probably a systemic problem, related to time lost in decision-making in a chain of command. To overcome such problems, commanders at every level should counterattack on their own initiative. They should have the means (not least the fire support) to do so. If the 22nd New Zealand Battalion had counterattacked earlier at Maleme Airfield, the Germans' pyrrhic victory on

Crete in 1941 might have been an even more costly defeat.[17] Best practice in both World Wars appears to have been to conduct immediate counterattacks from the lowest level to defeat the enemy's attack, together with deliberate counterattacks subsequently, if required.[18]

How should a defensive force be organized to harass, delay, disrupt and counterattack an enemy attack? This book is not a tactics manual, but some suggestions can be made. The key features would be: an outpost line to deceive, confuse, harass and delay the attacker; a series of strong points to fix him; counterattacking forces at platoon, company and higher levels; and fire support available to everybody from platoon commander upwards.

The balance, or emphasis, is critical. Suppose a battalion commander has three companies, each of nine sections in three platoons. He plans a defence with two forward companies dedicated to positional defence and his third company dedicated to counterattack. Eighteen of his 27 sections are in positional defence and nine available for counterattack. There is only one counterattack force. As an alternative, consider that only the forward sections in each forward company are in positional defence. All other troops are used for counterattacks. Only eight sections are now dedicated to positional defence, and the remaining 19 to counterattack. The emphasis has gone from $2/3$ to $1/3$ in favour of positional defence, to less than $1/3$ to $2/3$ in favour of the counterattack. There could be up to four platoon counterattacks, two at company level, and a battalion counterattack. The difference would be significant, and produce a far more flexible and elastic defence.

These infiltration concepts should be underpinned by three key mechanisms. The first is a mechanism for protection in the attack. The dissipation of tempo due to concern for one's flanks is a key failure in the attack. The case of the British 3rd Division on D-Day in Normandy is particularly galling, since it was predictable in advance.[19] The eastern flank of 3rd British Division (and hence the whole amphibious landing force) would lie on the double obstacle of the River Orne and the Caen Canal. The 6th Airborne Division was dropped immediately east of those obstacles to secure the lowest bridges over the canal and river (including the now-famous Pegasus Bridge). As the 3rd Division advanced south from the beaches, its developing left (east) flank would be covered by 6th Airborne Division. That aspect of the landing went as planned. The major enemy formation (21st Panzer Division) was encountered to the south, not the east. Nevertheless, at about 1300 hrs the corps and divisional commanders ordered the reserve brigade (8th Infantry) to deploy forces facing east behind the line of the Caen Canal, as a further line of defence behind 6th Airborne Division. The 3rd Division lost any remaining capacity to push southwards on D-Day. Perhaps surprisingly, the Germans' attack at Kursk in 1943 bogged down where the success of their own thrusts exposed flanks which they then had to protect, dissipating combat power from the spearhead.[20]

The answer appears to lie in *dynamic* protection. If each of a series of multiple attacks is conducted in column, the second echelon in each column can be tasked to protect the lead element's flanks and rear, as in Figure 15.

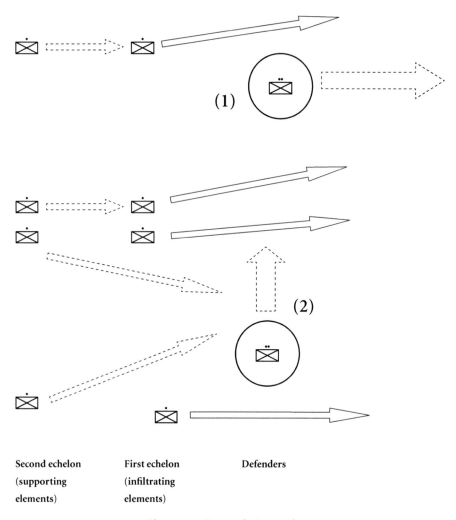

Second echelon	First echelon	Defenders
(supporting	(infiltrating	
elements)	elements)	

Figure 15: Dynamic Protection

Notes:
(1) This defending element resists until bypassed but then withdraws, under risk of being attacked in the rear.
(2) This defending element attempts to counterattack a bypassing attacker, but exposes himself to attack from flank and rear.

Dynamic protection has a major psychological benefit. Consider the commander of an attacking element. He has been ordered to penetrate rapidly. He might be given every encouragement to penetrate far into the enemy's depth. However, having made some progress, he might quite reasonably worry about his flanks. He would naturally be reluctant to penetrate too far unless he was confident that his flanks and rear will be protected; in this case, by the element behind him.

This should not appear at all surprising. General Israel Tal's division used the technique in the Six-Day War.[21] The US army employed it for armoured formations in 1944–45.[22] In a sense it was the epitome of medieval Swiss battle tactics. They typically used a three-unit formation. The vanguard marched straight towards a key point in the enemy's line and attacked it. The main body followed it, and attacked to one flank or other in support. A small reserve would attack either to relieve its colleagues or to secure victory.[23] Importantly, such tactics were essentially simple.

Patton also identified the process. He stated that '[a]gainst counterattacks, the offensive use of armour striking the flank is decisive. Hence a deep penetration by infantry, whose rear is protected by armor, is feasible and safe.'[24] The effect on the enemy should also be considered. Typically his position is ruptured, his forces widely and thoroughly suppressed, and he is probably also outnumbered overall. In such a position many enemy commanders will be disinclined to counterattack. If facing multiple penetrations, an enemy will have a hard job to choose which penetration to counter, and the decision-making environment will scarcely be benign.[25]

The second mechanism is speed. Guderian considered it 'perfectly tenable' that, 'having penetrated an opposing front, an attacker's best protection lay in continuous movement, since the enemy would be unable to execute effective countermeasures'.[26] Speed protects, confuses the enemy and pre-empts him.[27] For example, the speed of the US 6th Armored Division's breakout from Normandy prevented the enemy from coordinating a defence (except in coastal towns) and 'unquestionably the Division bypassed far greater strength than it attacked'.[28]

If tanks can move at 60 kph on roads, they can travel across country at 40 kph in some circumstances. This has important implications. If a tank company occupies about a square kilometre and moves at roughly 40 kph, a defender trying to hit its flank has a window of about 90 seconds. Too late, and he misses it; too early, and he loses surprise and might expose his own flank (see Figure 16).

This clearly presents the defender with a difficult task in coordination. I have seen commanders almost panic in a situation like this during an exercise.

The third mechanism is bypassing. It is sensible to drive a long way to avoid conflict, not least in order to maintain high tempo. Driving round the enemy also places one at his flanks or rear, or at least threatens to do so. It does mean that local flare-ups are to be anticipated as bypassed pockets react.[29] This is a sign of

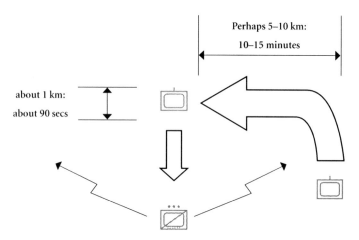

Figure 16: Combat at 40 kph

success, not failure. Bypassing one's own forces is another aspect. Staff colleges make much of the detailed staffwork required to pass one force through another in the attack. They are probably right to do so. The whole process is doubtless complex and time-consuming. One can attempt to do it at the lowest level (that is, arrange for a company to pass through another rather than a brigade through a brigade), which requires flexible regrouping at higher levels. The ability to regroup quickly is a key battlefield mechanism. Alternatively, one can avoid the problem by driving around the in-place force. As Patton (quoting Kipling) put it: 'fill the unforgiving minute with sixty seconds [worth] of distance run'.[30]

We have described a synthesis of encountering, infiltrating and exploiting, coupled with the use of indirect fire, to inflict shock and surprise. The elements should flow into each other. 'Encountering' is an insidiously mild term which should include aggressive ground reconnaissance to forestall, pre-empt, harass, disrupt and seek opportunities for infiltration and indirect fire. Infiltration techniques, mounted or dismounted, should be simple and flow naturally from the advance. The enemy's flanks and rear should be identified and exploited, demoralizing and surprising him and initiating his collapse. Further echelons should continue to exploit, again using simple techniques and making simple but timely decisions.

This process is not complete until the enemy is destroyed: rendered incapable at least for the duration of the engagement. That necessitates some damage, but extensive damage is inherently inefficient, and generally uneconomical in terms of logistics. We seek the destruction of the enemy's force, not his forces. That is, we seek a systemic effect. Large-scale surrender is a key indicator of success. Multiple infiltration coupled with stunning firepower and exploitation should have systemic effect, if incorporated into a coherent whole. The concept should include the extensive use of raiding (with air power, long-range artillery, reconnaissance and light infantry) to harass, disrupt and isolate the close battle. We have seen that surprise and aggressive use of ground reconnaissance are more effective than any likely range of force ratios, and we begin to see how. Airpower which reduces the effectiveness of enemy long-range sensors and interdicts the movement of his reserves has a similar effect. A combination of all three is a campaign-winning combination.

How should we assemble people and equipment in order to carry out those tactics? What shape and size should their organizations be? What mix of troop types they should contain? How should they be controlled? We have repeatedly stressed that we should focus on effectiveness, but how should we achieve that? When we assemble those people and equipment, we expect some synergy. How do we actually achieve that? Before looking at those questions we should compare armoured formations at the end of the Second World War and at the close of the twentieth century. The differences are illuminating.

A British armoured division of 1944 contained three armoured and four infantry battalions organized into two brigades. There were two field artillery battalions with a total of 48 guns, one anti-tank and one anti-aircraft battalion, and three engineer companies. There were three supply companies and one ordnance company. The division contained a total of 14,964 all ranks, 246 (cruiser) tanks and 1,453 trucks.[31] US armored divisions and German panzer divisions were surprisingly similar.

The British divisional HQ contained 51 officers. There was one brigadier (the artillery commander), one colonel (the divisional surgeon) and seven lieutenant-colonels.[32] Main HQ could operate on the move. German HQs were smaller (about 25 officers) and US HQs larger (79 officers).[33]

Divisions became smaller, more mobile and simpler during the Second World War. British, German and US armoured divisions all started the war with three brigades or regiments, and had one brigade or regiment deleted after one major campaign. The Germans deleted Panzer and Panzer Grenadier Brigade HQs,[34] and the Americans deleted regimental HQs, in both cases simplifying internal structure. This was probably due not just to resource constraints: within limits, smaller formations appear to be more effective. The British significantly augmented their armoured divisions in just one key respect. Between 1941 and 1944 tank numbers remained static, but truck numbers grew from 858 to 1,453 per division.[35]

By 1999 British, US and German armoured divisions had all become much larger. In the British case there were three brigades with a total of six armoured and six infantry battalions. There were three artillery battalions with a total of 96 guns; four engineer battalions; and ten supply and ordnance companies in two battalions. There was an anti-aircraft battalion and a helicopter battalion, but no anti-tank battalion. The division contained 348 tanks, about 25,000 all ranks and about 7,400 vehicles.[36]

The HQ contained about 160 officers. It contained no more brigadiers than in 1944, but seven more colonels and 11 more lieutenant-colonels. Whereas in 1944 engineers, supply, maintenance and ordnance were commanded through what was effectively a battalion HQ per service, there were now several battalions of each. There was effectively a brigade HQ each for engineers, maintenance and logistic support embedded within the division. There is therefore an extra layer of bureaucracy. American and German divisions were broadly similar. This enlargement seems to have been steady, incremental and to an extent uncontrolled. Divisions were not deliberately enlarged so much as augmented,[37] for reasons that appeared sensible at the time. As we shall see, the rationale was almost certainly less logical than it appeared. Armoured divisions were smaller in the Second World War because experience had shown that larger divisions were less effective.

Since convoy road speeds are still the same as in 1944, the division now takes almost three times longer to pass a point. Furthermore, despite (or perhaps because of) its greatly enlarged staff, the HQ now takes longer to produce orders. This is difficult to quantify, but in 1944 divisions typically gave orders one evening for the next day (i.e. about nine to ten hours from receipt to execution).[38] Patton considered that 'a division should have 12 hours and, better, 18 hours between the physical receipt of an order at divisional headquarters and the time it is to be enacted'.[39] Divisions now seem to plan on 36–48 hours and consider anything less somewhat hurried.[40]

Logistic support had become extremely complex by the late 1990s, with a plethora of specialist platoons and companies each supporting particular units or functions. This is in direct contrast to the findings of the British Army's Bartholomew Committee on the lessons of the 1940 campaign in France. That study specifically directed that all second- and third-line supply and transport companies be interchangeable so as to provide more flexibility in the control of logistics.[41] The Committee observed that 'much greater flexibility was now required, particularly in respect of third line transport, which was in fact never used in the way for which it had been designed'. A very similar observation was made in 2000.[42] Similarly, the Bartholomew Committee specifically sought 'to impose greater responsibility on junior officers and NCOs' in the supply units, whereas the ratio of officers to trucks in 2000 was almost exactly *twice* what it had been in 1944.[43,44]

Some armies are demonstrably more effective than others, and some of those differences are due to different organizations and tactics.[45] Years of seemingly well-intentioned organizational change appears to have had negative results.

Divisions take longer to produce orders, so C2 is actually slower. They take longer to move, so operational tempo is reduced. Their HQs are considerably more complex and bureaucratic, and probably less effective. The French army created semi-permanent divisions after the Seven Years' War (1756–63). They were in part intended to allow formations to march and fight as one fluid operation, eliminating much of the time needed to deploy before fighting. Modern divisions have seemingly lost that ability.

The size and shape of armoured divisions in 1944, the results of the Bartholomew Committee, and the size, shape and operating characteristics of divisional signals were available for study in the MoD Library and elsewhere. So was, for example, the report of the Hartgill Committee on army medical services of 1941.[46] I know scarcely anybody in the armed forces who has even heard of them, let alone studied them. In part, this reflects a failure to study military history in breadth and depth. Other reasons for this drift away from hard-won experience, and how to avoid it, will be considered later.

We should consider military organizations as fighting systems. We seek to generate a beneficial set of emergent properties. That is not easy, since those properties *are* emergent. We can, nevertheless, make some sensible observations. Beyond that, trial and experiment are needed. The primary aim is effectiveness. A lack of economy can be accepted within reason: we have already seen that the provision of reserves is, strictly, uneconomical but necessary. The real issue is 'how uneconomical?' Some inefficiency is desirable. Extreme efficiency is a consequence of close coordination, which is undesirable to some extent because it is susceptible to disruption.[47] War is complex and uncertain, and lies towards the edge of chaos. Disruption to closely coordinated processes should be expected.

We have also seen that forces should be flexible and agile. Articulation, subdivision into quasi-autonomous elements which cooperate within an overall purpose, aids flexibility and agility and is also robust.[48] 'Agile' suggests 'mobile' and 'responsive'. The mobility of a force is the result of two main factors: the types of vehicles and how many of them there are. The types of vehicles impact in several ways. Heavy-tracked vehicles have good cross-country mobility. However, when considering the mobility of the force overall they are limited. Tanks need transporters for long road moves, or the numbers available at the end of the march will be considerably reduced through breakdown.

The number of roads and tracks which heavy armoured vehicles can use in marginal terrain is another factor. For example, when the German army entered Kosovo in 1999 there was only one road, but seven other tracks or trails across the Albanian border. The 70-tonne tanks could only use the road, whereas (for example) British 8-tonne Scimitar reconnaissance vehicles could probably have used six or seven other tracks, considerably increasing the options available.[49]

In addition, columns containing tanks typically move at only 30 kph, whereas a fleet of (say) 8-tonne wheeled armoured vehicles could potentially travel at 60 kph or more.[50]

The number of vehicles in a force is the other factor in its mobility, due to the length of road columns. The time taken for the whole force to pass a point (the 'pass time') is a matter of hours for modern brigades and divisions, and can easily exceed the time for any one element to travel along the route (the 'run time'). Table 1 illustrates this effect, by considering four possible cases.

Table 1 Run Time, Pass Time and Total Move Time for a Formation

Case	Total Vehicles	Convoy Speed (kph)	Run Time (hrs)	Pass Time (hrs)	Total Move Time (hrs)
1	1,000	20	5.0	5.0	10.0
2		40	2.5	2.5	5.0
3	2,000	20	5.0	10.0	15.0
4		40	2.5	5.0	7.5

All cases assume a move of 100 km, with 100 metres between vehicles. In these cases the pass time is one-half to two-thirds of the total move time. This can be reduced by moving on multiple routes. For example, Case 3 would be reduced to 10 hours on two routes, or 8 hours 20 minutes on three. The figures given above are roughly illustrative for the brigades of the late twentieth century. A British mechanized brigade, with its artillery, engineer and logistic support, contained 1,532 vehicles in 1999.[51] Conversely, in 1944 a division contained only 2,500–3,000 vehicles.

Another paradox can now be resolved. Tanks can now move significantly faster than they did during the Second World War. Yet formations move no faster. In simple terms, if they are twice as large now their pass times will be twice as long. Hence any potential improvement in run time has been negated by increase in pass time. There are probably also other reasons: for example, formation HQs now take longer to produce orders.

The span of command also dominates the overall characteristics of a force. We shall use 'span of command' quite narrowly here: the number of combat groupings assigned to a commander. Thus in a 'square' brigade the span of command is four; in a triangular brigade, three. The cumulative effect of span is easy to visualize. Consider a force of a number of companies of a given size. If there are three companies per battalion, three battalions per brigade and three brigades in a division the total number of companies in the division is 27. If the division was truly square (four brigades each of four battalions each of four companies), the total would be 64. That is, about 2.4 times as many. Assuming that the number of vehicles in the division is in proportion, the pass time of the formation would increase 2.4 times as well.

Span of command is important in an entirely different area. There is a psychological limit to the number of subordinates one can control. Where their work interacts dynamically that number is typically five, plus or minus two. The number decreases under stress.[52] Combat is inherently stressful and arguably the greatest stress comes from the control of close combat rather than (say) indirect fire. The span of command for combat units or formations will tend to be low. Observed spans of command are typically three or four, plus an indirect fire commander and engineer and other specialists.

However, operational research suggests that the practical span of command is even lower. Trevor Dupuy analysed more than 200 battles and identified that the average number of subordinates *actually committed* to combat at any level from platoon to army was normally only 1.742.[53] This included organizations with a span of command of four, three and in some cases two. The figure of 1.742 is highly significant. If the number were as high as two, then one would expect an army to place 32 companies in its first echelons (32 is 2 raised to the power 5; that is, five layers of command). Yet 1.742 (the observed number) raised to the power 5 equals only 16.041 – that is, only about half as many companies in the first echelon of an army.

Thus armies structured into companies, battalions, brigades (or regiments) and divisions regularly commit rather less than two manoeuvre subordinates at every level, and the 'rather less than' is significant. Unless this deployment pattern changes significantly (which, given the above remarks about span of command, seems unlikely), manoeuvre forces can be seen as systems which routinely commit two or less manoeuvre subordinates to battle. We shall use the example of a division below. If we continue to see a division primarily as a manoeuvre formation (which we will revisit later), then it is a system to deploy and sustain at most two brigades, four battalions and eight companies in its first echelon. The organization of the division should be optimized to do that. If anything, the resulting formations will be larger than strictly necessary (since 2 is larger than 1.742).

A corporal commands a section of soldiers on a well-defined task using drills. A number of corporals and their sections grouped together is a platoon, and a number of platoons is a company. Drills are still highly significant at the company level in all armies, and the company is the highest level that can effectively be commanded by one man.[54] Although company sizes vary, they tend to be of one size for one branch of a given army. For example, all tank companies in the German army have the same number of tanks. Hence the company is a useful 'building-block' when discussing units and formations.

The key question is how big a division should be in order to sustain eight companies in its first echelon. Commanders at all levels should be able to concentrate force with combat support and combat service support, and have suitable reserves. Not least, they need suppressive firepower at all levels. At various levels they need to be able to scout for information, gain information through sources such as EW and reconnaissance aircraft, and conduct raids. The conceptual basis for this is to be able to deploy two manoeuvre elements forward, plus troops for reserves, and to form a main effort at each level, whilst checking for emergent characteristics.

The provision of reserves is a major issue. Most echelons of command contain a third and possibly a fourth subordinate of equal size as a reserve. This may be appropriate at some levels, but it quickly becomes uneconomical. If we design a division based on a span of command of three at all levels there would be 27 companies in the division; yet only eight or less are committed at any one time. Do we need 27 companies in order to sustain eight in combat?

The answer to this is no. A study of British and dominion infantry divisions (of nine infan-try battalions, hence 36 companies) in the Second World War shows that all nine battalions were never used on the same day by any of the six divisions in five battles (a total of 81 division days in battle).[55]

The divisions only ever used eight battalions on the first day of a major attack. They could presumably have used fewer if their parent corps had been differently organized. They only used seven battalions in similar circumstances, or in the case of very major German attacks. Even in the case of relatively static, positional infantry combat, divisions do not seem to need more than seven battalions. This is not conclusive, but it is insightful.

A further example is that of the 1st (UK) Armoured Division in the Gulf War.[56] The division contained 21 tank and infantry companies in six battlegroups. On no day were all six battlegroups employed. On every day at least one battlegroup HQ, two armoured squadrons and one armoured infantry company were not employed. Thus even in a formation that small, there were always more forces available than were used on any one day: let alone in any one engagement.[57]

In the Gulf War, the 1st (UK) Armoured Division was about the size of a Second World War armoured division. Due to the Rule of Four, however, the British 7th Armoured Brigade had grown to 16 companies by the Iraq War of 2003. In the same conflict within the US 3rd Infantry Division (an entirely mechanized formation) 18 companies sufficed for not one but two brigades. The Rule of Four had become a real impediment.

Table 2 Utilization of Battalions within British and Dominion Divisions, 1942–44

Division	Battle	Days Committed	Days in which battalions fought:								
			1 bn	2 bns	3 bns	4 bns	5bns	6 bns	7 bns	8 bns	9 bns
9th Australian Inf	El Alamein	10	10	–	–	2	6	1	1	–	–
78th Inf	Liri (1)	13	4	3	3	2	1	–	–	–	–
51st Highland	El Alamein	12	1	1	1	3	2	1	2	1	–
1st Inf	Anzio	23	7	6	5	2	–	2	1	–	–
15th Scottish	Epsom (2)	6	–	–	–	1	–	3	1	1	–
15th Scottish	Reichswald	17	1	5	6	2	1	2	–	–	–
Totals		81	13	15	15	12	10	9	5	2	0

Notes:

(1) Italy, 1944.
(2) Normandy, 1944.

We have already suggested that the bigger a force, the less mobile it is. In 1955 Field Marshal von Manstein was invited to comment on the proposed structure of the Bundeswehr prior to its establishment. For political reasons, West Germany was committed to establish 12 divisions. The proposal suggested armoured and mechanized divisions with strengths of 12,575 and 13,200 men respectively. The plan also included corps and army troops which Manstein calculated would add an average of five extra battalions to each division (two armoured and one each of artillery, infantry and reconnaissance). His comments included:

a. [with the addition of corps and army troops] the already oversized Divisions will become even more unwieldy or unmanageable ... Thus the divisions become more unwieldy than they already are.

b. The last war showed that motorised units of more than 12,000 troops were very unmanageable.

c. During the last war one saw repeatedly that, because of the excessive time needed to reach actions stations from their marching columns, and the inherent difficulty of leading them tactically, these large divisions had to be broken down into smaller fighting units.

d. ... the hitherto planned 6 tank and 6 infantry divisions are too large and too inflexible ...

e. The planned tank and infantry divisions are tactically and operationally unmanageable; on the one hand insufficiently mobile or manoeuvrable, and on the other insufficiently potent.[58]

Being constrained to 12 divisions, Manstein recommended the *enlargement* of divisions to about 25,000 men. This allowed the formation of very powerful brigades. Manstein explicitly stated that '[i]n reality the suggested brigades are equivalent to small divisions'.[59] Critically, the internal structure of those brigades contained regimental staffs between brigade and divisional level. Thus Manstein's proposed brigades can be seen as systems to generate and sustain up to eight companies in combat. They had the internal structure of a division.

The key issue is that of reserves. Combat is not fractal. The case for having three companies in a battalion, and probably three battalions in a brigade, is easily understood. However, at higher levels the assumption of three equal subordinates rapidly becomes uneconomical. A third brigade is a whole brigade that is not normally used. Given that German, British and US armoured divisions in the Second World War deployed only six–seven battalions, it appears that the extra battalions are not needed. The British example had only two brigade HQs. The US division had two large Combat Command HQs and a rudimentary third (a colonel's command with only three officers, five soldiers and no signallers.[60] Thus it would also appear that a third brigade HQ is unnecessary.

Such formations typically held only one battalion as divisional reserve. Smaller reserves can generally be committed faster than large ones. Committing reserves at higher levels (for example, corps rather than division) will also take longer than doing so at low level. All this tends to suggest employing reserves that are 'two down' above brigade level (i.e. a brigade at corps level, and a battalion at division). A typical panzer corps in 1944–45 contained two divisions and an independent regiment. Van Creveld remarked that the flexibility of a formation stands in reverse proportion to its size.[61] In 1944 the Wehrmacht reorganized all its infantry divisions from nine battalions to seven.[62] It appears that this was not simply a matter of resource constraint.

The counter-argument is that committing such reserves cannot be decisive, because they are too small. That would be relevant if it were simply a matter of numbers. Where speed and time are critical, however, the ability to get fresh troops fighting is probably more important than their numbers. This appears to have been the case in the Second World War. It is entirely consistent with the concepts of high-tempo, dynamic, surprising and shocking combat we have discussed. It appears that smaller, more agile formations are more effective than large ones. Our discussion will therefore proceed along the lines of about 18–24 companies in a division of six to eight battalions in two brigades. The organization of the Israeli Southern Command in the Sinai during the 1967 Six-Day War is a good example.

That campaign is one of the greatest examples of high-tempo armoured operations since 1945. The Israeli army shattered the numerically superior Egyptian forces, broke through the depth of the enemy position, and then exploited with breathtaking speed to the Suez Canal. It deployed a total of nine brigades in three divisions, hence ostensibly a span of command of three at division and corps level. There were a total of five armoured, two mechanized and two infantry brigades.[63] The campaign is illustrated in Figure 17.

Yehuda's brigade was committed solely to clearing the Gaza Strip. It collaborated with Tal's division, but was substantially independent. Kutti's infantry brigade formed part of Sharon's division for the break-in battle in the south. Thereafter it mopped up and took no part in the mobile battle. The main offensive westwards was conducted by six brigades in three divisions. All three divisions, plus the brigade in the Gaza Strip, attacked in a single echelon. Thus at Command (corps) level they were more than 'three up'. However, no division attacking westwards ever contained more than two brigades; brigades were repeatedly regrouped between divisions; and often a division had a brigade leading on one axis with only a single battlegroup on its second axis.

Thus the Israeli success was achieved with very small divisions, but rapid regrouping and very fluid C2 relationships. Had the Israelis chosen to operate with three triangular divisions, they would probably have committed four brigades in the first echelon. But it is unlikely that that would have generated the

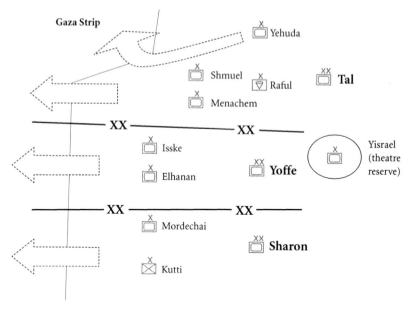

Figure 17: Initial Disposition of Israeli Brigades on the Southern Front in
the Six-Day War

same flexibility and tempo. In this instance rapid regrouping and very flexible C2
supported a high span of control at the theatre level, reflected by a narrow span
of control within divisions. This seems to be a very clear example of the agility
of small, highly mobile formations fighting autonomously.

This is an argument for smaller formations, not smaller forces. The argument
is that a force of the same size organized into a number of smaller divisions is
more effective than one of fewer, larger divisions. A 'seven-battalion' division may
be more effective than one of 12 or 16 battalions. It certainly seems no less so.
Two large divisions with a total of six brigades would, however, almost certainly
be more effective if reorganized into three smaller divisions. That would allow
the corps commander the option of committing three divisions to combat (hence
24 companies in the first echelon) rather than two (and just 16 companies).

Dupuy's research suggests that he probably would not do this. Conversely, the
Israelis did in 1967. Having the flexibility to do so could be critical. A computer
simulation conducted in DERA CDA in 2001 found that the three smaller
divisions were more effective for two reasons. Two small divisions initially
committed did as much damage to the enemy as two larger divisions. They
sustained fewer casualties in the process – because they presented a smaller, more
dispersed target to enemy long-range fire (and possibly because they were also
more agile). The key issue, however, was that an opportunity arose to introduce
the whole of the third division into a gap around the enemy's flank. The corps

commander probably did not have more than two divisions in combat at any instant, but the *possession* of a higher-level reserve was decisive in this case.[64]

The previous paragraphs suggest a core of manoeuvre forces that largely frame the size and shape of the division. The optimum size of an armoured division seems to be about 18–24 companies in six–eight battalions. The next issue is how combat support and combat service support (CSS) troops should be scaled and organized. Relevant factors include the need to sustain the force, the need to concentrate force and switch the main effort, and the deadening effect of overprovision.

Every commander within a close-combat force should be able to suppress the enemy, but this should be interpreted sensibly. Only those subordinates actually in contact need be supported at any one time, not every company and battalion throughout the force. It is normal practice to allocate artillery and mortar observers permanently to all combat companies and battalions. However, the guns only need support two committed brigades, hence four battalions and eight companies. The corps commander will normally only commit two divisions. The balance between dividing all guns equally between divisions and holding a large proportion at corps level seems to favour the latter. Why scale a division that spends a third of its time out of battle with a large number of guns? Better surely to give it relatively few, and hold a large number available to support the divisions actually committed. In the 1990s all the British Army's AS 90 self-propelled guns were grouped at 96 per division.[65] Conversely, in the 21st Army Group in 1944 there were only 48 or 54 25-pounder guns per divisions, but 38 battalions of artillery at corps, army and army group levels.[66]

How large should reconnaissance forces be? Armies appear to use reconnaissance platoons in all battalions that could have an independent role; hence they were absent in the Red Army until the end of the Cold War.[67] There is then an organizational issue. Brigades probably need their own reconnaissance companies if their role is substantially separate from the other brigade(s) in the division. In practice, brigade and battalion-level reconnaissance units have almost never existed together within a divisional framework. Divisions have generally owned a reconnaissance battalion of three or four companies, of which one could be allocated to a detached brigade if required.

In passing we should discuss the idea of 'aggressive reconnaissance'. It doesn't mean fighting for information. That merely results in destroyed reconnaissance vehicles. There is a tension between the requirement to gather information by stealth and the need to exploit it aggressively. There is also a need for guard forces which fight to pre-empt, to seize fleeting opportunities, or delay. Both aspects suggest the need for genuinely all-arms groupings, of which reconnaissance vehicles form but one part.

Anti-tank defence illustrates the need to take a systemic view of combat and of armed forces. Anti-tank weapons destroy tanks, whereas anti-tank troops protect units and formations. All combat troops need to be able to destroy tanks. There is a case for giving support troops anti-tank weapons as well. There is, however, a wider issue: the need to protect the unit or formation. The Germans appear to have realized this, and made two deductions not made in other Western armies. The first is that 'protection' and 'defence' are not the same; anti-tank units can be handled offensively and be very effective. There are numerous examples drawn from the western Desert.[68] Lieutenant-General Lesley McNair, who organized and trained the US army for the Second World War, insisted that anti-tank battalions should be handled aggressively, and then refined their tactics after experience in Tunisia.[69] The second deduction is that there is a strong case for a formation-level specialist anti-tank unit that fights as such. In the Wehrmacht it was equipped and commanded as a battalion, with an armoured headquarters.[70]

The counter-argument is that 'the best defence against a tank is a tank', which is just not true. Specialist anti-tank weapons are generally about 2.5 times better at destroying tanks.[71] Guderian freely admitted it.[72] During the Battle of the Bulge US tank-destroyer battalions destroyed 306 German tanks; one battalion alone destroyed 105.[73] Perhaps more importantly, the argument that 'the best defence against a tank is a tank' has been used by the armoured community to justify replacing anti-tank units with more tanks. The US army is one example.[74] Western armies seem to have accepted the argument that a tank unit is more flexible than an anti-tank unit without adequate analysis. There is a systemic need to protect the force, but in practice tanks are almost never used for anti-tank defence. They are usually given more offensive roles, such as counterattacking. Tying them to the defence is seen to distract from their offensive role. The culture of armies' armoured branches gets in the way here.

Should a division have an anti-tank battalion or another battalion of tanks? Von Manstein's specific, and perhaps surprising, response was to want the anti-tank unit.[75] This whole issue is confused in English-speaking armies eyes by a failure to understand the difference between German assault gun and anti-tank units. Suffice it to say that Wehrmacht divisions contained both, even where the equipment in use was practically identical.[76]

Combat engineers are an interesting case. There is without doubt a case for a company with each brigade in combat. There may be a case for permanently assigning one to every brigade, so that they can work together in peacetime. There is often a need for more engineers, for example when preparing defended positions or for obstacle-crossing operations. Unfortunately, engineer equipment is bulky and slow-moving: consider not just the armoured bridge-layers, but also the bridge-transporters (like tank-transporters) which carry spare bridges. One can easily reach the point where the sheer number of engineers and the size of their equipment is an impediment to the formation's agility. In 1999 a British armoured brigade typically contained 1,532 vehicles, of which each battlegroup typically contained 140. The brigade engineer battalion had 139, including all of the largest vehicles.[77]

The basic requirement is to support two manoeuvre subordinates in contact and form a main effort. In this case an engineer company per (forward) brigade, with a third within the division, would seem sufficient. That force could be commanded by a lieutenant-colonel with a battalion staff. In 1944, it was.[78] In the 1990s a British division had four battalions commanded by a colonel with brigade-level staff.[79]

The same can be said of almost all CSS components. A division needs enough trucks to supply its two forward brigades with enough to fight with. The C2 arrangements required to coordinate demand with supply and movement usually requires one company per brigade. Artillery ammunition is typically 85 per cent of a formation's logistic requirement,[80] so further companies are needed for that. The rest of the division's requirement is small (less than about 5 per cent), and can be met from the companies already described. Thus our division might need three to four truck companies and, again, this could be commanded by a lieutenant-colonel with a battalion staff. Again, in 1944, it was.[81] By 1999 there were two large battalions of logistic support troops, commanded by a colonel with a brigade-level staff in a British division. Medical and maintenance support is generally similar. German and US practice in the 1990s was also similar. Placing the bulk of the artillery at corps or higher level would mean that the large number of trucks needed to support them would not reduce the mobility of the divisions.

In all such cases (reconnaissance, aviation, engineer, supply; or medical or maintenance) there may be a need for one 'reserve' company to form or support the main effort in our small division. That is far more economical than a structure of several battalions plus a brigade staff per service. By analogy, there may be a possible requirement for an extra supply, maintenance and medical battalion at corps level.

Modern-day raiding troops consist primarily of attack helicopters and armoured reconnaissance units. The range of attack helicopters and the long planning time needed for deep penetration missions suggest that they, like aggressive ground reconnaissance, are best considered as deep operations forces at the higher tactical or even operational levels. This tends to suggest that divisions should not conduct deep operations using attack helicopters, long-range aggressive reconnaissance and missile systems. Those troop types should be held at the theatre level, although there may be a case for grouping attack helicopters at divisional level for close air support missions.

Detailed calculations based on the foregoing discussion suggest that a 1990s 'NATO' division, which typically contained about 25,000 men and about 7,500 vehicles, would be better divided into two of about 12–13,000 men and 3,500–4,000 vehicles.[82] The smaller divisions would have 18–24 companies in six–eight battlegroups, as opposed to about 48 companies in 12 battlegroups. They would have a total of about 18 battalion-sized units, as opposed to about 36.

Our simple model of complexity also suggests that the current, larger division is considerably more complex and hence more difficult to manage. We have used the parameter of $\frac{n^2}{2}$ as a measure of complexity. The relative complexity of the two divisions is therefore about

$$\frac{36^2}{18^2}$$

or about 4:1. This is another indication of why staffs have in practice grown since the Second World War. Their divisions are more difficult to control.

This discussion of organization has been based on the assumption that divisions are primarily manoeuvre formations. It has followed the argument that they should be relatively small in order to be effective. Effectiveness is considered in terms of the ability to inflict shock, surprise and suppression, using highly agile and mobile close-combat forces. A different approach would be to substantially reduce the emphasis on shock action. This might see the ground manoeuvre component reduced to one brigade, and a significant 'raiding' component introduced (armoured reconnaissance battalions with tanks, AH battalions and possibly light infantry for heliborne raids). The emphasis could shift yet further, with the shock action of close combat eliminated, and replaced with light or mechanized infantry solely for base protection and possibly for static defence to set the conditions for raids. However, such a formation would not be able to conduct shock action as currently understood. The risk is that in the absence of shock action, combat tends to be indecisive. Such a formation could not conduct the functions of a division as we currently recognize them.

The main purpose of this discussion of structures has not been simply to propose smaller divisions; rather it has been to suggest that we have been looking at the issue the wrong way, and that theory and history both suggest better alternatives.

7

Commanding the Battle

Command in battle appears to be largely a matter of deciding and acting faster than the enemy in a complex and dynamic environment. That begs two questions. The first is how best to make decisions in combat. To answer that we shall consider the characteristics of combat in turn, and make deductions for decision-making. The second question is how to organize the resources needed to support the decision-maker and turn his decisions into action.

We now see combat as highly dynamic, complex and lethal; based on individual and collective human behaviour; waged between organizations which are themselves complex; and, fundamentally, not determined. What can we deduce from that about tactical decision-making?

First, combat is highly dynamic. The life of a soldier, the survival of a tank or the impact of a salvo of shells can turn (decisively and fatally) on the actions of a matter of seconds. Many decisions must be taken quickly, and many some of them will have life-or-death consequences. As we will see, there is significant advantage to be gained by making decisions materially faster than the enemy. That increases the need for speed. Without being specific at this point, it is clear that commanders must be able to make tactical decisions very quickly, in an absolute sense.

Secondly, combat is complex – in the extreme. It is fundamentally a detailed interaction of a large number of agents and factors. Each agent can act, and react, in many ways. It is impossible to predict the precise outcome of any set of actions. Battlefield decision-making must deal with a huge number of inter-related factors. The term 'deal with' is apposite. There are at least two distinct approaches. *Considering* every individual factor, in detail, if it is possible, is one option. An alternative is to attempt to 'deal with' the complexity of the problem as a whole.

Combat is lethal. That has two implications. Firstly, commanders must continually seek to meet their assigned goals, and inflict damage on their opponents, while protecting themselves and their own forces. That is a major constraint. Secondly, decision-making will frequently take place under conditions of fear and stress. The result will often appear to be less than strictly rational. That is not a criticism: it is a necessary consequence. Human behaviour should be the starting-point. Tactical decisions are made by human beings. The behaviour of the human brain is the key.

Combat is based on individual and collective human behaviour. Historically and anecdotally, men have known a great deal about how soldiers behave in battle. Sometimes they achieve great feats against great odds. Sometimes they run away. At times they stand their ground doggedly, and occasionally die to the last man. Sometimes they are slaughtered like sheep. Anecdotally and historically, soldiers know that they should exploit the strengths of human behaviour and avoid its weaknesses. Battlefield decision-making should do the same. These things have always been known, but more recently the human sciences have begun to address the relevant aspects of behaviour. We now know more about how soldiers will tend to behave in given circumstances, so we should be able to exploit that knowledge.

Armed forces are complex institutions, and no two are the same. Managing an organization as complex as a division, or even a platoon, requires that the whole organization be motivated effectively to the immediate task; not least because its survival, let alone the achievement of its goals, depend on it. This 'motivation to the task' requires effective communication of intent from the commander to all subordinates. The key word is 'effective'. The degree to which all individuals in an organization act to support the overall direction is one marker of the efficiency of the organization.[1]

Combat is not simply 'of' complex organizations, but between them. Extremely complex interactions take place between opposing forces in combat. Some of those interactions will be particularly strong. The battle may wage to and fro as each side attempts to gain the advantage. Death, incapacitation, gain and loss of ground and collateral damage will result from those interactions. Thus battlefield decision-making must cater for strong interaction on the battlefield. This is not confined to interaction between opponents. The idea of synergy suggests that the units of an army can cooperate to improve their net effectiveness. That is, there can be a strong interaction between units on the same side. The corollary is that if one element decides badly, nothing useful happens – or worse.

Unfortunately combat is not determined. Thus we know that there may well be strong interaction, but we cannot know with accuracy what that interaction will be. The implication is simple: if we cannot know what the outcome will be, we should guard against undesirable outcomes (especially since combat can be fatal). We should also be prepared to exploit unexpected benefits. If we view combat as a highly complex and almost chaotic system, decision-making should be robust against rogue outcomes.[2] That requires both the provision and retention of reserves, and the mental and physical flexibility to be able to react as the outcome unfolds.

Thus it seems that tactical decision-making should be very quick. It must 'deal with' many interrelated factors. It should aim to inflict damage whilst avoiding damage to one's own forces. It should exploit the strengths and weaknesses of the human beings involved in combat, both friendly and enemy. It is often undertaken in highly stressful circumstances, not least the fear of death or dismemberment. It should initiate, and accommodate the outcomes of, strong interactions between forces on the battlefield, be robust against rogue outcomes of those interactions and yet support the clear communication of intent from commanders to subordinates throughout the chain of command.

This is a relatively long list of requirements, and some appear to be mutually inconsistent. The need for speed mitigates against a careful consideration of the many factors involved. The strong interaction between battlefield agents suggests detailed study and modelling, yet such models almost inevitably rely on deterministic rules. Unfortunately war does not appear to be susceptible to determinism. Instead of trying to specify precise activities and outcomes, we should specify general intentions and desired outcomes: the 'sort of thing' we wish to result. Trying to simulate the complexity of the interactions involved would require vast computing power. Such computing power currently requires many hours of real time to analyse a few moments of simulation, which militates against supporting tactical decisions in real time. If recent trends continue, advances in computing power will be harnessed either into more detailed (and similarly time-consuming) simulations, or in quicker simulations which are considerable simplifications of reality. Thus, whilst it can be insightful, computer simulation is unlikely to produce great breakthroughs in tactical decision-making in the early twenty-first century.

This excursion into computer power highlights the general problem. Combat is too complex for detailed, highly analytic decision-making – not least because there is generally not enough time to make highly detailed plans. Yet some commanders have made successful battlefield decisions and have done so frequently. Some armies have produced commanders who have done so repeatedly. Decision-making is a human process, and some humans in some circumstances can make good battlefield decisions. Sometimes they have done so very well. The solution appears to be to consider how real human beings make decisions under stress.

There is compelling evidence that the quantity of information used by battlefield commanders to make battlefield decisions is in practice very small. Much of the information gathered is not considered in the making of decisions for which it is requested.[3] As a corollary, the effort expended in collecting information is commonly out of proportion to its utility.[4] The consequences of such unthinking quests for information are considerable.[5] In practice, battlefield commanders rarely make decisions according to the highly structured methodologies taught in Staff Colleges.[6] We shall consider this later. Critically, battlefield decision-making is not information-*intensive*; it is information-*sensitive*.[7] The difference is critical. It has major implications for the function and organization of an HQ, and many other issues.

It is lucky that commanders do, and should, use very little information in making tactical decisions, since there is significant battlefield advantage in making and disseminating decisions very quickly. To understand why requires a simple model. Imagine two opponents, A and B. Assume, firstly, that both are equally likely to make good battlefield decisions; and secondly that both take equally long to do so. Both start at the same time. There are four possible outcomes:

- Both A and B make good decisions. Their decision-making gives neither any advantage.
- Both make poor decisions, with similar results.
- A makes a good decision, and B makes a poor decision; or
- vice versa.

In either of the latter two cases there will be a clear advantage. Unfortunately, since they are equally likely to make a good decision, the chances of A or B winning are equal. Their decision-making gives neither a clear advantage, other than by chance. These outcomes are illustrated by the decision tree at Figure 18.

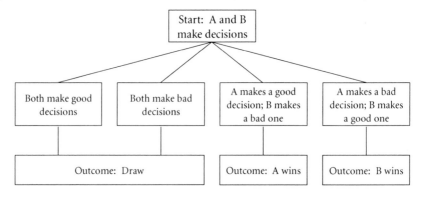

Figure 18: Initial Decision Tree

The outcome is dominated by the likelihood of a draw, and this is relatively insensitive to the adversaries' decision-making abilities. To illustrate this, consider firstly that both A and B have an 80 per cent chance of making a good decision. The probabilities of either A or B winning are then 16 per cent, while the probability of a draw is 68 per cent. Conversely, if A has only a 60 per cent chance of making a good decision and B a 95 per cent chance, the overall probability of A winning is 3 per cent; of B winning is 38 per cent; but the chance of a draw is still 59 per cent.

Now consider the case where B has the capacity to make a very good decision – one with a high probability of creating a battle-winning advantage. In this case, however, A can make a decision and turn it into action twice as fast. Once again there are four initial possibilities:

1. B makes a good decision. However, A makes one in half the time, pre-empts B and wins the engagement. A's decision-making created a battle-winning advantage.

2. again makes a good decision, but A makes a bad one in half the time. There are then two subordinate outcomes:
   If A then makes a good second decision, the result will be a draw.
   However, if A makes a bad second decision, B will win.

3. A makes a bad decision, but so does B. A gets a second chance. If he decides well, he wins. If not, neither does (that is, a draw).

4. B makes a bad decision, but A makes a good one. A wins.

Then there are six possibilities in total. These are shown on the decision tree in Figure 19.

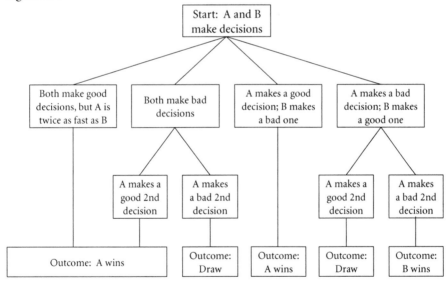

Figure 19: Modified Decision Tree

The outcomes do, however, depend on how good A and B are relative to each other. As an illustration: if A has an 80 per cent chance of making a good decision, but B, taking twice as long, has a 95 per cent chance:

- The overall probability of A winning is 80.8 per cent.
- The overall probability of B winning is 3.8 per cent.
- The probability of a draw is 15.4 per cent.

That is to say, despite being less likely to make a good decision, A's ability to decide twice as fast has created a clear advantage. The likelihood of A winning in this case is almost exactly equal to the probability that he will make a good decision, whereas the chance of B winning is almost nil.

This simple model hides two second-order effects. Firstly, where A initially makes a poor decision, he can learn from the situation. There is probably a better chance of him making a good second decision, which improves his chances overall. Secondly, B will realize that A has pre-empted him. The situation for which he is planning is changing; he may be demoralized by A's success, and the chance of him making a good decision is probably reduced.

The model is simplistic, since it only represents decision-making at one level. In combat many commanders, at many different levels, make interlocking decisions on both sides. War is a highly complex interaction of both physical and psychological factors. Psychological aspects are far more complex than the two second-order effects of learning and pre-emption. Shock, surprise and stress are all major factors. Nor does the model suggest what a 'good' decision is. However, accepting those limitations, it does suggest some advantage to a force which systematically makes decisions faster than its opposition. We have used a speed of 'twice as fast' to make the computation simple, but the figures correspond well with the Soviet finding that a force that can react twice as fast can defeat one five times as large.[8] It also helps explain the Soviet emphasis on speed of decision-making.

The Soviet norm for a reinforced tank or motor rifle battalion mounting an attack from first contact was 25–60 minutes.[9] A German battalion in the Second World War was expected to conduct an attack from the march within 40 minutes of first contact.[10] Figures for performance at formation level are hard to come by. In the Second World War, General Sir Miles Dempsey, who commanded the British Second Army in Normandy, considered that an infantry battalion given four hours to prepare for an attack on a prepared defensive position would have a 100 per cent chance of success.[11] Many Western armies consider that an HQ should take no more than $^1/_3$ of the total time available for planning, leaving $^2/_3$ for subordinate echelons. Applying the '$^1/_3$; $^2/_3$ rule' to General Dempsey's figures implies six hours for a brigade and nine hours for a division. The British 11th Armoured Division conducted a 30 km night march to capture Amiens on the night of 30–31 August 1944. H-Hour was six hours and 45 minutes after the division was first warned of the operation.[12] The 2nd SS Panzer Corps' order to counter the Arnhem landings was given verbally within an hour and a half of the first British paratroopers landing. A written order followed within three hours.[13, 14] The operation order was never changed, and from the German perspective the outcome was entirely successful.

Nowadays, the 'going rate' for a trained British battlegroup is an hour –

50 per cent longer than the Wehrmacht or Soviet examples. In the 1990s the HQ of the Third (UK) Division estimated 12 hours for the production of an operation order on change of mission.[15] The US army's *FM 101–5* quoted nine hours.[16] At best, this implies 27 ($= 9 \times 3$) hours for the planning and execution of a divisional mission. War is allegedly more complex today than during the Second World War. But is it really? In the Second World War, formations of all arms and services were controlled in near-real time; with formation HQs commanding three to four major subordinates (units or formations); and typically interacting with similarly organized enemy. That has not changed. We should be doubtful that any increase in the apparent complexity of modern war is justified by the increase in staff numbers, or that it justifies the *decreased* operational tempo that has resulted. We will discuss this later on.

Our model also suggests that tactical decisions, and operation orders, do not have to be as good if they are produced significantly faster. In the previous section we considered a case in which A had an 80 per cent likelihood of making a good decision, in half the time that B needed to have a 95 per cent chance of making a good decision. Let us now consider what happens if A makes a weaker decision – say, only 70 per cent likely to be good; or 60 per cent. Figure 20 shows how the three outcomes vary with A's chance of making a good decision.

The graph assumes that B has the same high chance of making a good decision as previously (95 per cent). A could afford to make decisions that are objectively quite poor. Even if he has only a 40 per cent chance of being correct, he can still create *some* advantage over B. If he has a 50 per cent chance, he is over twice as likely to win as B (51 per cent versus 23 per cent)! The driving factor is the probability of A making a good decision. The more likely it is that A will make a good decision, the more likely it is that he will win; and the less likely B will win. At higher probabilities (55 per cent and above), it is also less likely that a draw will result.

Clearly this is only illustrative. The model is naïve, as we have discussed. The results for A deciding, say, three times as fast would be even more marked. The key is not the objective quality of decision-making but its speed. There is often an implicit assumption that a good operation order, reflecting a good tactical decision, is a detailed one. This is not so. A good decision is one which is 'about right', but made and turned into action much faster than the opposition does. Objectively that is much faster than at present. As Patton put it: 'A good solution applied with vigour *now* is better than a perfect solution ten minutes later.'[17]

We can now see that highly structured estimates which attempt to consider all the factors involved in combat are limited. One study calculated the time needed to analyse all the factors used in a typical Staff College estimate rigorously, and concluded that such analysis simply does not take place.[18] 'Dealing with' complexity can now be seen to be a perfectly understandable, if

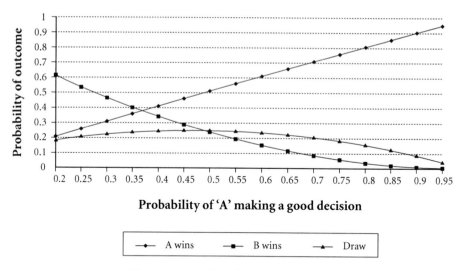

Probability of 'A' making a good decision

Figure 20: Quality of Decision-making

largely subconscious, process of forming a mental model of the problem and envisaging a solution that will work. The resulting plan is a relatively high-level abstraction. High-level abstraction is a workable way of dealing with complexity. It can now be seen to be entirely appropriate for tactical decision-making. The result is *not* a closely coupled, detailed plan: it is a general idea, a 'sort of thing' which the commander can envisage as working.

Decision-making which generates a general idea, a 'sort of thing', can be done very quickly. Structured, analytic decision-making cannot. Even trained and experienced Staff College graduates require hours to make detailed analytical decisions. This is due at least in part to the relatively slow process of capturing the various factors and deductions in oral or written form.[19] Yet a quick guess will not usually do. The effectiveness of tactical decision-making seems to follow a hump-backed curve: virtually no benefit for no time input, followed by rapidly increasing benefit as an appropriate amount of time is spent to make a quick but considered decision. That decision produces a decision which is 'about right' and encapsulates little more than the right 'sort of thing'.[20] As more time is spent on producing a fuller, more structured decision, the result may be more precise and detailed, but the benefit is little (if any) greater. In fact, the benefit falls to zero if the time taken becomes so great that the enemy can make a good-enough decision faster.

Tactical decisions will often be taken under conditions of fear and stress. Fear and stress reduce the effectiveness of mental processes, not least decision-making. However, we now have a model of decision-making in which sub-optimal decisions are acceptable, as long as they are taken quickly. That is not to suggest that decisions taken under stress are better than those made in less stressful conditions. However, it appears that experienced decision-makers can accommodate stress in a way that structured decision-making methods cannot.[21] The effect of stress is probably more pronounced on higher, more conscious mental processes such as analytic decision-making than on the subconscious processes which experienced decision-makers use.

One can consider decision-making as a process which incurs overheads. Time is one. In a perfect world, HQs always foresee developments in the situation, make timely decisions and relay those to the troops so that they can move seamlessly from one activity to the next. History tells us that this often does not happen. The reasons are not those of time delays in message transmission. Since the advent of the telephone, message transmission has occurred at the speed of light. It cannot get any faster. Delays mostly occur within staff processes.[22] Where there are several echelons of command, this can take many hours.

Time is one overhead. Bureaucracy is another, in two respects. The first is the number of staff and their support personnel: the signallers, drivers, cooks and clerks involved in command and control. A Western army typically employs about 600 people to support a single divisional commander.[23] There is an implicit assumption that that overhead is beneficial, because the force is more effective as a consequence. That assumption is flawed: we will see that modern Western armies are top-heavy.

Bureaucracy is also a costly overhead in terms of precision.[24] Decisions generated by large staffs tend to be mediocre and reflect sociological issues such as consensus: 'lowest common denominator' tactics, perhaps. Decision-making structures, and processes, which allow decisions to be taken locally wherever possible, are more likely to result in effective decisions, all other things being equal. But all other things are, of course, often not equal. The commander on site may react under stress in a way which he would not were he more detached, for example. However, workable processes are available to mitigate the worst consequences of local perspectives.

The second main issue for this chapter is to consider how we should organize the resources needed to support the decision-maker, and turn his decisions into action. What we see on the battlefield is, broadly, a series of tactical command posts (CPs) distributed throughout a deployed force. What *do* they do, what *should* they do and how might they be improved? *Ad-hoc*, ill-trained HQs are dangerous and have resulted in major tactical failure and loss of life. Therefore this issue is important. Not least, the higher the level of the CP, the greater its ability to influence the outcome of battle, positively or negatively.

Much of this section is based on first-hand observation, both on my part and that of a number of psychologists and analysts with whom I have worked. I did not observe any incompetent or stupid staff. Everybody I saw was able and well intentioned. What follows is a criticism of organizations, systems and processes, not of individuals. However, the same problems occur in many places – across a range of HQs. Thus the problems we see are fundamental and not unique to any one HQ.

Evidence, including that from the fighting phase of the Iraq War of 2003, suggests that our HQs have become too large; contain too many overlapping functions; have officers of inappropriately high rank; plan too much; and tend to be very busy. Yet their activity is not matched by their output, and they issue orders which are too long and complex and arrive too late.[25]

Deployed HQs have become unwieldy. During Operation TELIC in 2003, the HQ of 7th Armoured Brigade was established for 42 officers but actually contained 96. That was in a total of 383 all ranks, excluding the signal company cooks, drivers, etc. The HQs of the two British armoured brigades in Operation GRANBY in 1991 contained only 288 and 306 personnel respectively. This growth of 25 per cent in 12 years is not accounted for by changed functions. Analysis of staff posts exposed unnecessary duplication and unwarranted growth. The trend for formation HQs to grow has been observed in several places. Recent operational analysis indicates that in a typical formation HQ, 40 per cent of the staff do nothing useful, and a further 20 per cent produce output that is largely nugatory. Thus only 40 per cent produce useful output.[26] Overall it appears that much of the apparent complexity of modern war stems in practice from the self-imposed complexity of modern HQs.

In 1974, HQ 20th Armoured Brigade reviewed its CP structure explicitly to provide a 'lean, hard, flexible and survivable Brigade HQ'.[27] The result totalled 105 all ranks and 30 vehicles of all kinds. It did not contain many of the functions required of a brigade HQ during Operation TELIC. Adding those officers, and soldiers pro rata, would bring the total for HQ 7th Armoured Brigade in Operation TELIC to about 166 all ranks and 48 vehicles. Furthermore, tactical CPs are now too large and too slow. Anecdotal evidence from Operation GRANBY spoke of CPs trailing the battle, out of touch.

The overall impression of CPs in Iraq in 2003 is of HQs that were large and usually very busy, but which produced relatively little output. Despite its augmented size, HQ 7th Armoured Brigade provided only eight written orders in the 18 days between 21 March and 6 April. In the same period the divisional HQ produced 27 written orders, but nine contained only miscellaneous coordinating detail. In several cases the orders resulting from this process were excessively long. On one 25-page operation order, the mission first appeared on page 10! A battalion 2IC reported that his unit HQ had produced an operation order one inch thick prior to G-Day. However, an hour after the beginning of the operation only one page was still relevant. During the Cold War, brigade orders rarely exceeded ten pages plus annexes, not least due to the physical difficulty of transmission. (There were no laptop computers during the Cold War.)

Much of this criticism would not affect operational effectiveness directly. It would simply keep excessive numbers of staff officers busy. However, the critical impact on Operation TELIC was that, on important occasions, the relevant orders were released too late. For example, five orders needed before H-Hour were only released by the British Divisional HQ the day after operations started. Operations to enter Basra are another example. The city fell early on the morning of 6 April. HQ 7th Armoured Brigade rushed out an operation order at 10 a.m. that day. It acknowledged that some of the events in the order might already have taken place. They had! The divisional HQ rushed out a fragmentary order, which said very little of substance, two hours later. Nor is this unique to the British Army: a staff officer in the 1st US Marine Division commented that 'The planning cycle was way behind the execution being conducted by the forward commanders. Div HQ was still producing lengthy [orders] that were too late for the commanders, as they had already stepped off.'[28]

The working rank in a divisional HQ in the British Army has normally been major and captain, and captain at brigade level. There has been a gradual trend since the late 1980s to place lieutenant-colonels in staff positions at divisional and even brigade HQs, and several majors into brigade HQs. The most serious effect is a tendency to overplan, since these higher-ranking staff tend to be planners rather than run current operations. It reduces the role of majors and captains at times almost to that of tea-boys. Lieutenant-colonels do not perform the same functions as captains and majors, but tend to require them as subordinates, increasing overall numbers.

The human aspect should not be underestimated. On one large exercise in which we looked at five different headquarters, we saw six significant failings of control in nine 'missions' or tactical phases. Only one was due to technical problems. All the rest were caused by human, not technical, factors. In another exercise I saw the intelligence branch invent a major enemy thrust which didn't exist, entirely due to their prior expectation as to where the enemy would attack. If I hadn't seen that I simply wouldn't have believed it. Nevertheless it happened, and when you see things like that first-hand you come to understand how critical mistakes actually happen in battle. The reasons are almost always human error. One of my colleagues estimates that most CPs make one major error roughly every 36 hours.[29] Command problems are more common than we suspect, and technical communications failures are a relatively small proportion of the total.

There are many *valid* reasons for the current size of HQs, but apparently no *good* reasons. Unchecked growth and lack of oversight is one. The growth of HQs since 1944 has been insidious, with no major step change.[30] Even quite major changes in requirements (such as NBC or 24-hour operation) produced only modest enlargements. The root cause appears to be that no body or organization has had oversight of the issue over a number of years.

Underpinning this whole issue is a major fallacy. It is that more information leads to better decisions and therefore greater battlefield success. That is simply not true. Not least, at some stage the information swamps the staff, producing information overload. As we shall see, adding more staff and more computers typically doesn't help. What is needed are decisions which are 'about right, but timely'. You don't need much information to make decisions that are 'about right'. You need the critical information, and you need it in a very timely fashion.

Furthermore, psychological research strongly suggests that the quality and quantity of information available have very little effect on the outcome of decision-making. Overall, the information available affects about 40 per cent of the content of a decision. Varying the level of uncertainty within the information provided has little impact. Three things drive the outcome of decision-making: the self-confidence and decisiveness of the decision-maker; the ability of the staff team to make sense of the data they have; and the social cohesion of that team. The experience of the decision-maker is also important.

This finding is devastating to conventional thinking on command. Many digital command systems are based on the premise that providing more and better information reduces uncertainty. It may do; but that has very little impact on decision-making. It implies that the billions of dollars applied to procure digital communication and information systems misses the target. The news is not all bad. Although quality and quantity of information only accounts for perhaps 40 per cent of the outcome, some information is high-quality and allows the commander to do better things. For example, knowledge of the location of the enemy's HQ may allow it to be attacked. Clearly it could not be attacked if its location was unknown. Yet even here human issues are important; a good or experienced commander and staff will know that enemy HQ locations are important, and direct sensors to look for them. What counts is not just luck nor sheer quantity of data, but skill and experience.

A further indictment of the impact of the digital revolution comes from historical analysis. A subtle and little-known study looked at improvements in command effectiveness, year by year, through the twentieth century.[31] After discounting issues such as weapons improvements, the study did find step-changes in the effectiveness of command. Unfortunately, the major step-change for most armies did not come between the two World Wars, the period when radio was generally introduced for tactical command. That improvement occurred in the later years of the First World War, when radio was almost totally unavailable. In practice the observed improvements came largely from organizational and procedural changes during wartime, not technology. It is too early to say whether the same thing will happen with the adoption of digital technology in the late twentieth and early twenty-first centuries. The evidence to date is that land tactical command has got worse, not better, since the end of the Cold War.

In November 2004 I spent some time with the headquarters of an Indian infantry division. The experience was salutary. Its structure was almost exactly unchanged from that of a British division of 1945, but the technology was roughly that of a NATO division of the 1980s. What was illuminating was that the staff seemed to be able to make decisions and act just as fast as a Western division could in 1945. Not four times slower, as we see in Western divisions today. Yet the staff felt that their CP was too big! Clearly that division could not do some things that a modern Western division can. But we really should ask whether or not all the technology and extra staff has generated any substantive progress.

A deployed land force contains many CPs. Since some CPs can contain hundreds of people, computers and communications devices, the resulting system is unutterably complex. This is particularly true if one also considers individual variation between people (for example, all brigade chiefs of staff are different) and the data contained within each CP. That data also changes constantly as new information is received, orders given and actions taken. Figure 21 is a gross simplification of this system. It focuses on just two aspects, the social and the technical. It is based on the premise that command systems are networks of people who communicate over distances using technical means. The upper layer shows units, CPs, sensors or weapons: groups of people who are co-located. The lower layer shows clusters of information systems at each location, and the electronic links between them. Social links are shown between the elements of the upper layer. When deployed on operations, those social interactions largely take place using the technical means in the lower layer.

CPs are also significant human artefacts and have social and cultural significance. They are representative of the commander and thus have totemic importance. They contain staff officers for whom a tour of duty there has important career implications. They are critical command nodes and thus subject to enemy attack. During the Cold War, NATO headquarters typically moved every 12 or 24 hours to minimize the risk of enemy artillery fire. That imposed a premium on small size and mobility. CPs are meeting-places and hence sources of informal information, rumour and gossip. These social and cultural issues should not be overlooked. They explain much about how CPs perform.

Free-play field training exercises demonstrate most forcibly that this system is just not geared for war at 40 kph. Current thinking and processes just can't cope.

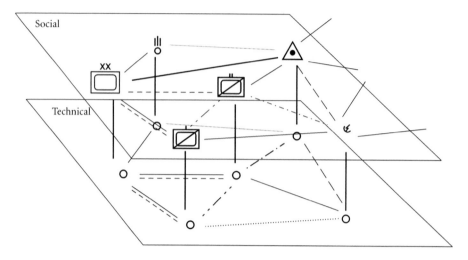

Figure 21: Command as a Socio-technical Complex

An enemy tank column is quite capable of travelling at 40 kph or so, even across country, if it is not actively stopped. That means it can travel 10 km in 15 minutes. I have seen enemy movement at that speed cause paralysis and disbelief in a formation CP on exercise. It required almost heroic moral and intellectual effort to rectify the situation. The CP staff almost suffered from shock. Procedures developed over years of peacetime CP exercises simply do not work fast enough. We should reduce and simplify procedures, and shortcut decision loops – mainly by empowering more junior commanders. Thinking and acting at 40 kph or more is well within the capability of our people, if appropriately trained. Witness naval warfare: a frigate can move at 30 knots; that is 60 kph. Ships are 'fought' by two Principal Warfare Officers (PWOs) and the captain – himself normally a former PWO. PWOs are produced by a specialist training course. Armies probably need such a course for staff involved in current operations.

CPs should be organized to make decisions that are 'about right but timely', and we have just seen that 'timely' might mean far faster than current arrangements allow. If the underlying decision should be 'about right but very quick', then the plan should not contain much detail. It is not much more than a concept. Trying to expand the basic decision into a closely synchronized plan is a fundamental error[32] and a pervading weakness – one of attempting to foresee the future rather than impose one's will on the enemy.[33] Combat is an astonishingly complex environment; attempting to impose a coordinated schedule on such complexity is folly. Methodologies based on prediction and control will inevitably be second best.[34] Commanders on the spot should be allowed to make and execute decisions based on the real situation, not on a closely synchronized plan made in advance[35] or at higher level. Furthermore, overcontrol tends to demotivate subordinates, and collective performance decreases.[36]

It is entirely wrong to attempt to extend such decisions into closely synchronized plans, with considerable detail coordinated in advance. That is to attempt a precision which the nature of the subject does not admit.[37] This presents an apparent difficulty: armed forces are astonishingly complex, and appear to require detailed coordination. Yet combat does not admit of closely coordinated plans. However, we observe that some armies can be, and have been, peculiarly effective in combat. This difficulty clearly can be resolved in a way that does not required detailed coordination from above.

A further error is that of trying to plan too far ahead. Combat is unpredictable. No plan survives contact with the enemy. Thus the situation after the next battle or engagement cannot be predicted with any accuracy. Planning too far ahead will normally be nugatory. If you plan to fight tomorrow, do not plan the next day in anything but the most general detail. You will clog up your communications systems, distract subordinate commanders and tire your staff unnecessarily. And if you are only going to plan one day ahead, why on earth would you want to spend more than a few hours doing so – at any level?

In general, we should design CPs to support commanders in seizing and retaining the initiative, imposing their will on the enemy and thereby winning. Armies don't get paid to come second, and no one would thank a CP which complained that the enemy moved too soon or too fast.

The orders which these CPs produce should be composed, and passed either verbally or in writing, as efficiently and effectively as possible. There is a considerable body of psychological and sociological knowledge about conveying intent and meaning. That knowledge supports passing simple, short, clear directives to subordinates.[38] Patton said that '[p]lans and orders should be simple and clear'.[39] An 'austere' message describing the situation and intended action is easy to convey and understand. It is also easy to check that the message has been understood.

The product of tactical decisions are operation orders that should be, can be, and have been, very short. They should be short for two reasons. Firstly, if the decision contains very little information, it is not sensible to tie that into a long and closely synchronized order. That is to endow it with a complexity which the decision itself did not contain, and which fundamentally restricts subordinates' freedom of action. Secondly, simple orders can convey simple intent clearly. The more words that are used, the more potential there is for ambiguity. Clearly there is a balance to be struck; the question is where the balance should lie.

In 1907 the German officer Colonel Helmuth von Spohn wrote that every instruction within an order constrains the subordinate commander's freedom of action. A decentralized command style such as Mission Command, which appears to be appropriate to land combat, therefore requires as few of them as possible. Operation orders therefore should be short.

In real, dynamic situations effective orders have often been very short. Guderian's order for the Meuse crossing in May 1940 ran to 36 sentences on 88 lines.[40] Although some aspects of that operation had been rehearsed, the order was typical of all Wehrmacht higher formation orders given in that campaign,[41] and possibly the entire war.

There is a perception that long orders are needed today (perhaps 30 pages for an 18-hour battlegroup mission, or 90 pages for a division)[42] because war is allegedly more complex today. We should be very dubious, for two reasons. Firstly, complexity is broadly a function of the number of elements in a system, the relationships between them and their dynamism. Thus on most counts the Second World War was much more complex than anything we see today. Secondly, how can we possibly know? War is, literally, unutterably and inconceivably complex. How could we tell if something is more complex than that? The broad thrust of this book is that attempting to deconstruct such complexity is futile. War may seem more complex today, but that is partly because we use far more information than we need to conduct operations.

Several conversations with commanders and senior staff officers indicate that what is needed in an order is: a very brief summary of the situation to describe why the mission is required; the mission; the concept of operations (what the commander intends to do to achieve the mission); missions for subordinates; and critical coordinating instructions. Most of the rest simply isn't needed, or should be passed routinely, separately from the order. For example, the instruction for handling prisoners of war for Operation TELIC did need to be passed, but didn't need to be a part of the main operation order.

However, what constitutes 'critical coordinating instructions' varies considerably. In one operation it may be the movement table; in another the artillery fireplan. This detail is often tabular and may be best produced as an annex to the main order. Identifying and producing that detail, ensuring that it is both critical and precise, is an issue of skill and experience. Some staffs produce a lot of detail because they think it is all needed. Such staffs lack the perception, skill or experience to work out what is and what is not needed. As a result orders get longer and longer, and the unnecessary detail constrains subordinates' freedom of action. There is a dichotomy between the requirement for making and enunciating decisions that are about right but very fast on the one hand, and precision of critical detail on the other. Establishing that balance is an entirely human issue. Aspects of personality, training and experience will dictate how well it is achieved.

It should now be obvious that it is impossible to synchronize a battle. At best you can coordinate it. Synchronization is the correlation of activities *at stated times*. Coordination is the correlation of activities *with each other*. Once one side has initiated combat, both sides lose the ability to dictate timings in advance. Activities may be pre-empted or delayed by the enemy. Western doctrine often includes the use of a 'Synchronization Matrix'. Its use was imposed on the British Army in the mid 1990s by a senior officer by decree without any analysis of its value, or the cost of producing it. The word 'synchronization' is taken to be more sophisticated, and therefore better, than 'coordination'. It seems to be that simple, and such cultural issues shouldn't be underestimated. Some armies seem to have a penchant for long orders simply to demonstrate effort or apparent 'complete-ness'. Cultural issues such as this contradict the need to give direction which is 'about right but timely'.

Patton was in the habit of writing orders for 3rd US Army on one piece of paper. In fact the order was typically contained on one side, allowing the other for a sketch or map.[43] The order for the French Deuxieme division blindé (Second Armoured Division) to liberate Paris in 1944 ran to one page. It seems that once a formation is thoroughly trained, it doesn't need long orders. In peacetime the emphasis should be on the training, not long orders.

Given such training, it is possible to take almost any operation order and reduce the essential detail to one page, supported by a small number of annexes. I once did so to a ten-page brigade order to the complete satisfaction of one of the battlegroup commanders, who went on to become a major-general. I have reduced several orders to one page since then. The vast majority of the detail is either not needed, repetitious, or should be reduced to standard operating procedures. Thus orders *can* be very short. As another example, the British Army's operation order for the annual 'marching season' in Northern Ireland used to be about 50 pages long. One year a colleague of mine reduced it to 12 pages. The main order was only two pages long. The rest was annexes.[44]

It seems that far too much process has been invented, and applied in inappropri-ate places, over 50 years of peacetime. Little of it has any objective value. British examples are 'Targeting Boards', the Intelligence Preparation of the Battlefield and the production of the Synchronization Matrix. These processes result in products which research indicates are virtually never updated, nor referred to by units in contact. There is no evidence that they ever *will* or *should* be.

What is needed is for a set of staff processes which are fast, effective and efficient. The CP should monitor the outside world continually, respond to changes swiftly and produce direction that is concise, effective and correct in its critical details. The CP is not alone. Its processes should be nested within those of the others in the hierarchy because there are systemic effects. Consider a situation in which a division takes nine hours to plan an operation. Its brigades might need six hours and its battalions four. Very simply, if all those steps take place consecutively the overall process would take 19 hours. However, given a high degree of concurrency, the whole process might be completed in 10–12 hours. Clearly the division's plan must be completed before the brigades can finalize theirs, and likewise for the battalions, so it will take more than just nine hours.

There are established ways of planning in parallel down a chain of command, and digital technology supports some of them. Observation from the Iraq War indicates that we still can't do things as fast as in the Second World War, so computers aren't the whole answer. One officer who participated in chain planning in Iraq described it as 'tortuous', adding that '[it] leads you up a lot of blind paths' as the superior headquarters explores options that don't result in part of the plan. Planning for such options is more difficult for junior headquarters with smaller staffs, so this is a real problem. Some of these procedural issues still require development, and old-fashioned military virtues such as self-discipline and leadership play a part. One of the reasons why a formation could react as fast as it did in the Second World War, despite the level of technology, was that the formation as a whole had streamlined its processes at a system-of-systems level. This suggests that much more emphasis should be placed on training, and less on technology.

In current doctrine, divisional and brigade HQs should plan for the next operation, whilst units conduct the current one. It is sensible, where possible, to plan not just for the intended next operation (the 'sequel') but also some alternatives ('branches'). Some eventualities could take place at any time; so contingency plans should be considered to cover them. Contingency planning provides both mental rehearsal and a sharing of intent. Map exercises have many of the same benefits. Given that the real circumstances are not likely to be predicted accurately, generating extensive plans based on contingency planning will normally be nugatory. A short fragmentary plan giving only the outline of the contingency, possible missions and key coordinating detail should be sufficient. That plan might occupy one side of paper, together with a sketch.

Decisions need to be made quickly, and they need to be 'about right'. Therefore decision-makers need appropriate information, and that information must be timely. It must be received and acted upon before the situation has changed and rendered the plan irrelevant. We have seen that the information needed is a fairly high-level précis or abstraction of the situation. Its content should be less detailed at higher levels of decision-making. Critically, the bulk of the detail is simply not required, especially if gathering and abstracting the relevant précis is time-consuming. What is emphatically *not* needed is a large HQ which can handle vast amounts of data and plough through it, imposing time delays.

Consider a tank regiment of about 100 tanks moving along a route. It is perfectly possible to report the location, identity and characteristics of every single tank. That would constitute a large amount of data. Yet for many purposes a report akin to Figure 22 would be all that is needed. Because the tank regiment icon is a single symbol on many military computers, Figure 22 is almost trivially small in terms of data. The benefit of sending such a report around a network, with the original detail available from its source only if required, is considerable. All deployed military communications networks have limited bandwidth, and large amounts of data clogs them up very quickly. In addition, the ability of the receiving CP to read and understand it quickly is much enhanced.

At 101434Z NOV 08

Figure 22: Tank Regiment Pictogram

Earlier we suggested that battlefield commanders rarely make decisions according to the highly structured methodologies (such as the Estimate Process) taught in Staff Colleges.[45] Such methodologies may even hinder expert decision-makers. The rigorous generation and consideration of alternative courses of action required by such methodologies is often a charade.[46] However, in some manner, information *is* fed to commanders; and decisions *are* made.[47] Most decision-making by battlefield commanders appears to be based on recognition of the situation as being typical of a class of problem to which the decision-maker can perceive a satisfactory outcome.[48] Such real-life or 'naturalistic' decision-making is quite different from 'analytic', structured methodologies favoured in business management courses and typified by the Estimate Process.

Experienced decision-makers appear to be able to make very good decisions with remarkably little information about the situation. The mental picture, or simulation, which decision-makers use is based on very few factors (typically no more than three) and a very small number of changes of state – five to six.[49] The brain rarely handles more than about five or six pieces of information simultaneously.[50] The amount of information actually used in making a decision is probably no larger. Evidence to support this comes from an analysis of juries in murder trials, who consider at most 10 per cent of the facts presented during the trial in forming a verdict.[51] However, the five or six pieces of information are usually high-level abstractions.

This recognitional or 'naturalistic' decision-making should not be seen as a substitute for structured, analytic procedures, but as an improvement on them.[52] In many combat situations such decisions are more likely to have a positive outcome than the products of highly structured processes. Thus battlefield commanders often do, *and should*, make good decisions using naturalistic processes. They need very little information with which to do so. However, the information used is entirely dependent on the particular situation.

The American psychologist Gary Klein studied many decision-makers under stressful and time-critical conditions.[53] His general finding is that novice decision-makers use structured methodologies, but experienced decision-makers generally do not. They make better decisions, and make them faster. As soon as the novice has some insight into the typical problem (the conduct of a platoon attack, for example) he tends to change from overt, structured methodologies to implicit, recognitional methods. In such methods, consideration of the problem prompts recognition that it is a type of problem with which he is familiar. He can envisage a solution which will probably work, although it may not be the best possible solution. This process is neither black magic, divine inspiration nor guesswork. It is how a real brain works.

The business studies section of most public libraries contain several books on decision-making. The great majority promote structured, analytic decisions. Many of those methods are highly mathematical. They are not suited to tactical decision-making because of the time required to conduct them properly.[54] Recognitional or naturalistic decision-making seems to be appropriate in many battlefield circumstances, but we cannot expect the commander (particularly the novice commander) simply to guess. Explicit methodologies are still needed. They help novice commanders learn, and serve as a fall-back for unfamiliar circumstances.

What seems to be required is decision-making which is about right but very quick; decisions which are made largely subconsciously; and small expert staffs to support the decision-makers. Naturalistic decision-making should be the preferred mode for experienced commanders and staffs when faced with a broadly familiar situation. It should not be mandatory. However, major Western armies appear to believe that decision-making should be highly structured, explicit and require large staffs. Culture does, by definition, persist. Thus there seems to be a cultural barrier here which should be explored and subsequently overcome. A skiing analogy is useful here.

To a novice, skiing is difficult. Several linked difficulties present themselves. One is gravity. The novice's instinct tells him to lean backwards, that is *uphill*, to avoid falling over. However, with tuition and practice, he may find skiing easy. Critically, the skier learns that leaning forward, generally *downhill*, improves performance dramatically. A very experienced ski instructor once told me that 'in 28 years I never told anyone to lean backwards'.

This analogy is quite powerful. Leaning downhill, 'off the slope', is appropriate because skiing is a dynamic activity. To be able to steer, and hence control, one needs to apply force to the front of the skis, and the only appropriate way to do that is to put weight on them. Since one generally skis downhill, that means leaning downhill. The critical link with combat decision-making is that skiing, like combat, is dynamic. Were it not, there would be no need to steer and it would not be sensible to lean 'off the slope'. It would be tiring and the skier would often fall down. However, skiing is inherently dynamic. There is no point in skiing if you don't intend to move. Combat is also inherently dynamic.

Combat is dynamic but much peacetime training is not – or, more accurately, is inadequately dynamic. Rarely adversarial, and with decision-making and staff procedures taught as discrete processes in classrooms, the dynamic nature of combat is rarely perceived short of war. In peacetime a soldier's experience tells him that decision-making should be highly structured, explicit and require large staffs. The culture of peacetime Western armies reflects that. The equivalent of 'leaning off the slope' is to control less; but with greater emphasis on speed, timeliness and precision in the few critical details. Field Marshal Lord Carver, who had commanded an armoured brigade during the Second World War, once said that in war a divisional commander had to think only as much as 30 minutes ahead. He also said that the British Army introduced command and staff trainers in the 1960s in order to practise tactical decision-making under stress.[55] That appears to have been forgotten.

We should employ the best brains in small groups, rather than try to assemble a collective brain. The trained human brain is a superb instrument. It is the best tool armies have to integrate information and formulate plans. Armies have established mechanisms to select and train the best people to do that. Given the right information and the right conditions, such people can come up with good plans quite quickly. You often see them when you look at a good unit or formation headquarters. Other people are clearly required to assemble the required information, and to translate plans into action. Well-trained small groups of such people can work very swiftly and efficiently. Interpersonal issues do, however, rapidly come to dominate the outcome. Strategies that give authority and responsibility to highly skilled individuals, or small groups of such individuals, seem to be the most effective. The worst solution seems to be trying to plan by committee. Unfortunately that is what you often observe in a CP. I once saw a unit HQ trying to plan with a team of 19 people. The unit contained only about 40 officers in total. Unit-level planning can be done perfectly well by about three or four people if they have the right skills and experience.

There must be some redundancy. Small teams can be very efficient, but staff become tired and start making mistakes; or they have to deploy away from the CP; or they may become casualties. They have to sleep sometime. One or two people available to support a team of three or four is still far more effective than a team of 15 or 20, in this kind of environment. Those team members must be generalists who can work across a variety of staff disciplines at the appropriate level of detail. That is one reason why armies have Staff Colleges; were it not for that, officers would receive training only in their own branch or service. The appropriate level of detail is not necessarily great, if the headquarters is working at the correct level of abstraction and not getting stuck in the detail.

Individual subject matter experts are sometimes needed, but that is not a reason to have them in every CP. For example, in April 2003 the main British HQ in Iraq needed expertise in blood supply and helicopter engineering. In both cases it was available within units deployed in theatre, and was used. The theatre medical and maintenance staff were generalists. They needed to know what questions to ask, who might know the answers, and be able to access those people. Psychological research indicates that teams of generalists are normally more resilient than groups of specialists.

Thus, in general, what is required are clear and logical processes by which a CP monitors the tactical situation, makes swift and effective decisions, and issues simple and clear orders in a very timely manner; then repeats that process whenever required. The CP also needs to pass information up, down and sideways as its contribution to the overall system described in Figure 21 (*see* p. 142).

Brook's Law relates to computer software engineering, but gives useful insight into the problems of organizing CPs. Fred Brook worked for IBM and became professor of computer science at the University of North Carolina. He observed that adding software writers to a team didn't necessarily speed things up. In practice they often slowed down. Software writing, like many intellectual activities, requires the people involved to interact. The time spent interacting rapidly becomes the dominant factor. This results in a 'bathtub curve' like that shown in Figure 23.

The graph shows how the time taken to complete a task (shown on the vertical axis) varies with the number of people sharing the task. Figure 23 shows the overall curve as the sum of two others. One curve shows the effect of job division. If a job takes, say, ten hours and it can be divided between two people then overall it will take five hours; with three people, three hours 20 minutes, and so on. This effect tails off as the number becomes large. So adding more people doesn't achieve very much. That assumes that the job can be divided perfectly and there is no need for interaction. The second curve shows the effect of interaction. Assuming that all team members interact with each other, the time needed to do so starts to rise rapidly. It soon begins to rise almost with the square of the number of people involved. This soon dominates the total time required to get things done.

Studies going back to the 1970s consistently indicate that when staff numbers are reduced, the effectiveness of an HQ improves.[56] All the evidence I have read suggests that CPs in Western armies tend to lie on the right side of the curve. They are already too big. Reducing staff numbers would increase the speed with which they could get things done.

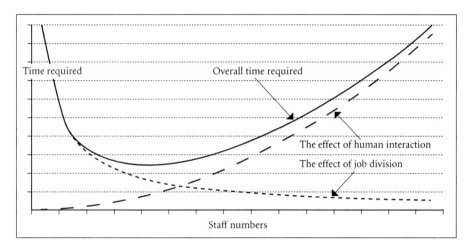

Figure 23: Brook's Law

Brook's Law is simplistic when applied here. Rarely does every team member have to interact with all the others. The slope of the curves and hence the location of the bottom of the bathtub can be redrawn many ways; notice that there are no numbers on Figure 23. It is, however, indicative. Unfortunately the effect is counter-intuitive. If a particular branch or cell is very busy, the normal response is to add more staff. When aggregated over many years since 1945, this effect contributes to large, slow and cumbersome CPs.

Many of the senior commanders and staff officers I have interviewed remarked that staffs should be small: rarely as many as 50 all ranks, even at corps level. The number is not related to the size of the operation. Small CPs are good: one general said that 'too large would have been disastrous' due to increased information requirement and managerial overhead. Commanders should learn to live without physical luxury or copious information. Some risk should be taken in information provision in order to maintain a small CP.

Using generalists rather than specialists is not the whole answer. The first curve in Figure 23 shows that if we move towards smaller staffs then each member's share of the work increases. Can people do that? There are two components to the overall task: intellectual work and human interaction. Human interaction is usually easier than the intellectual work described here. So we need people who can work well under time pressure and interact efficiently, plus the conditions which allow them to do that.

Human beings generally show a range of ability of about 5/6:1 for a given task.[57] That is to say, the very best can do the task five or six times better than the very worst. Figure 23 suggests that if each individual could work just 50 or 60 per cent harder we could move a very long way towards smaller headquarters. Again, the numbers are only indicative. In a team of staff officers you don't normally find people five or six times better than others. You do, however, see people who are noticeably better. That suggests there is a considerable and largely unrecognized premium on staff selection and training.

The conditions are also important. CP layouts can be a nightmare. Since the end of the Cold War they have tended to be organized so that everyone can see the same maps and charts. In so far as that supports shared situational awareness, that is fine. But the crucial requirement is to allow small, expert teams to work with minimal interruption. Unfortunately, current arrangements tend to guarantee interruption. A colonel visiting one headquarters during the Iraq War spoke of the effect of the need for almost continuous updates, which severely disrupted planning. It was also telling to observe, during a series of exercises, that the CP which had the best technical information systems also had the worst collective situational awareness. Digital technology is highly useful if, and only if, properly applied.

It is also quite obvious that almost everybody above the working level (captain and major) in an HQ is superfluous, except the commander and COS. This applies in almost every HQ, *especially* the very large ones. What the higher ranks achieve is a requirement for more briefings to them and meetings between themselves. During one four-day exercise which I observed, cell chiefs (who were lieutenant-colonels) spent 80 per cent of their time at briefings and meetings. For the most part such meetings merely ratify decisions that are obvious to junior staff. Worse, they sometimes make decisions which subordinates know to be flawed, but cannot reverse. It also produces an observable tendency for the commander not to be told the objective truth, but rather what senior staff officers believe he wishes to hear. This is dangerous, to say the least. A current British brigade HQ (with a major as COS) is just about OK, but rank representation is creeping upwards. Rank representation within a divisional HQ has grown alarmingly since 1945.

The critical point seems to come at the stage where more than one or two lieutenant-colonels are working routinely in a tactical headquarters. At that stage one of two things, or both, can happen. The upper staff echelons may begin to develop their own dynamics, as just described. Alternatively, junior staff become marginalized and therefore disaffected and typically ineffective. The problem is not the colonels themselves, but rather the fact that they are one or two ranks higher than the bulk of the staff, and that they normally have junior staff working for them (which further increases the numbers, and brings out the effects of Brook's Law). This is clearly a matter of balance. However, one can see how there is a premium on having small numbers of high-quality, well-trained and junior generalists in a CP.

What, then, is the ideal CP for land tactical operations? It should contain small numbers of selected and well-trained staff. They should be generalists not specialists, and they should be quite junior. The emphasis should be on quality, not rank. Specialist advisers should be available if required, but not necessarily present on the headquarters' establishment. Formation CPs need to work continuously around the clock for days or weeks at a time, which requires at least two or possibly three shifts of some functions. Three shifts of (say) five or six good generalists is much easier to provide than two shifts of (say) 20 specialists. If all these principles are followed rigorously, and digital information systems applied sensibly, one could easily see a division run by perhaps 20 staff in total and a corps by 30 or 40. Training those staff would be a key requirement. Much more investment should be placed in training them, both individually and collectively.

There are two simple but overarching conclusions. Firstly, the future is not digital: it's human. What is needed is things that bind talents together as a team, not more bandwidth. Secondly, given time, resources, open minds and not much money, we could revolutionize land tactical command. The key problems are human, cultural and institutional. What we need is a change of mindset, less overt process, far fewer people (generally in more junior ranks) and better training of some elements of the staff.

8

The Soul of an Army

We have considered what sort of tactics to employ, and how to organize and command forces for best effect. Since combat is fundamentally a human activity, we should also examine how individuals' actions contribute to operational effectiveness.

Many books have been written about military 'geniuses': great commanders who appear to stand apart from others of their time. However, few writers can agree why a particular commander was so effective. Few even agree as to who the geniuses are. Genius does not just happen; it seems to be an extreme combination of human characteristics which should in principle be observable and measurable. What kind of people have those characteristics? What kind of person would make a good or bad commander?

The nature of combat is unique; nothing else resembles it closely. Therefore successful commanders will probably either have long experience from a young age, or a predisposition to study, or both. Wellington, Wolfe, Alexander, Napoleon, Patton and Slim all fit this category in one way or another.[1]

Because the patterns and dynamics of combat are high-level and abstract – rarely can one actually *see* an armoured division in combat, except on a map – successful modern commanders are likely to be intuitive thinkers. In this sense 'intuition' refers to the presentation of information to the conscious from the subconscious. 'Intuitives' tend to think that way, unlike 'sensors', who tend instead to perceive the external world directly. Since the information is presented from the brain rather than the senses, intuitives are more prone to abstract conceptual thinking. Rommel was definitely an intuitive thinker. As a child he had a strong liking for abstract mathematical puzzles rather than childlike games.[2]

Intuition linked to both opportunity and a tendency to reflect is likely to result in learning from experience, be it one's own or others'. The key indicator appears to be the tendency to reflect, and is often revealed in written works. The books of Patton, Manstein, Guderian and Rommel all show this tendency.

War is about human beings, and people respond to leadership. The confident, charismatic air of a stereotypical leader normally demonstrates extroversion, considered as the tendency to draw emotional energy from the outside world, and hence one's fellows. The converse trait is introversion: introverts draw their emotional energy from within their own minds. Slim was doubtless an extrovert, as was Patton. Montgomery was probably not. Although Montgomery was popular with his troops, this may have been because he was successful, and carefully stage-managed his public appearances. He does not seem to have been particularly popular with his peers and immediate subordinates.[3]

Personality traits can be masked, particularly by intelligent people. Intelligent introverts can learn to demonstrate some of the more outgoing social conventions. Manstein may well have done so. Relatively high intelligence, as measured by IQ, does seem to be a marker of successful modern commanders. This is probably because understanding high-level abstract trends requires a certain intellectual horsepower.

We shall follow human typology later. We should, however, make a few remarks about bad commanders, since their alleged prevalence fuels an industry in 'great military blunders' or 'military disasters'. The early days of the Boer War were not crowned with success for the British Army. When he assumed command, Lord Roberts sacked five generals, six brigade commanders and 20 colonels.[4] That was a major proportion of all the British commanders in theatre. The BEF of 1940 had 17 formation commanders: the Commander-in-Chief, three corps commanders and 13 divisional commanders.[5] Their subsequent careers are described in Table 3. Where 'no further trace' is shown the officer retired, resigned or was discharged between the Fall of France and April 1941.

All were relatively distinguished: of the 44 UK-based division commanders in 1940, five had a Military Cross (MC), 17 had a Distinguished Service Order (DSO), and 16 had both the MC and DSO. One had a Victoria Cross *and* the MC and DSO! Only five had no decorations for valour or distinguished service.[6] Yet five of the BEF commanders immediately disappeared without trace and six never commanded field formations again. One was captured. Of the remaining five, three became field marshals, one commanded a corps and one another division. Thus in round terms one-third disappeared, one-third were re-employed away from the front line and one-third went on. This process was not limited to the army: the Royal Navy had about 140 captains in seagoing posts in 1939, but within a year only 40 were still at sea.

The statistics for the British Army in the First World War are at first sight less dramatic. However, due to of the huge expansion of the army, many pre-war officers were promoted rapidly early in the war. Many were in fact sacked.[7] Overall, it appears that a major proportion of those senior officers appointed to command the British Army as a consequence of its peacetime promotion system were found wanting in war. If culture is 'that which persists' in the behaviour of human institutions, one can imagine that such effects still occur. There is some anecdotal evidence that it does, and trenchant comments have been made about Edwardian values and attitudes persisting at least into the 1990s.[8]

Table 3 Formation Commanders of the BEF in 1940

Ser.	Appointment	Rank	Name	Subsequent Career
1	C.-in-C. BEF	Gen.	Viscount Gort	Commanded Gibraltar Fortress, Malta, then Palestine and Transjordan as a field marshal.
2	C.-in-C. I Corps	Lt-Gen.	M.G.H. Barker	No further trace.
3	GOC 1 Div.	Maj-Gen.	H.R. Alexander	Replaced Barker as corps commander before 4 June. Supreme Allied Commander Mediterranean as a field marshal.
4	GOC 2 Div.	Maj-Gen.	H.C. Loyd	COS Home Command then GOC London district.
5	GOC 48 (South Mids) Div.	Maj-Gen.	A.F.A.N. Thorne	Eventually GOC Scotland as a lt-gen.
6	C.-in-C. II Corps	Lt-Gen.	A.F. Brooke	CIGS as a field marshal.
7	GOC 3 Div.	Maj-Gen.	B.L. Montgomery	C.-in-C. 21st Army Group as a field marshal.
8	GOC 4 Div.	Maj-Gen.	D.G. Johnson VC	No further trace.
9	GOC 50 (Northumbria) Div.	Maj-Gen.	G. le Q. Martel	Director, RAC. Commanded 7th Armoured Division for a short time.
10	C.-in-C. III Corps	Lt-Gen.	R.F. Adam	Became C.-in-C., UK Northern Command then the army's Adjutant-General.
11	GOC 42 (East Lancs) Div.	Maj-Gen.	W.G. Holmes	Corps commander in the Middle East as lt-gen.

(continued)

Ser.	Appointment	Rank	Name	Subsequent Career
12	GOC 44 (Home Counties) Div.	Maj-Gen.	E.A. Osborne	No further trace.
13	GOC 5 Div. (GHQ Res.)	Maj-Gen.	H.E. Franklyn	Corps commander in the UK Home Forces, GOC Northern Ireland, then C.-in-C. UK Home Forces.
14	GOC 5 Div. (GHQ Res.)	Maj-Gen.	W.N. Herbert	No further trace.
15	GOC 46 (North Mids and West Riding) Div.	Maj-Gen.	H.O. Curtis	DC of Hampshire and Dorset until 1944.
16	GOC 12 (Eastern) Div.	Maj-Gen.	R.L. Petre	No further trace.
17	GOC 51 (Highland) Div. (in Maginot Line)	Maj-Gen.	V.M. Fortune	Captured at St Valery (subsequently died in captivity).

Conversely it appears that the Wehrmacht produced large numbers of good generals, from a peacetime base of only 3,000 officers before 1933.[9] During 1939–45 a total of 2,344 generals and field marshals served in the Wehrmacht. The Wehrmacht continued to produce capable generals throughout the War: few were particularly young, even in 1945.[10] About 500 were killed in action.[11] This is vastly different from the Allied forces. Only one British divisional commander was killed in action in north-west Europe during 1944–45[12] and only two US generals died due to enemy action during the Second World War, although Lieutenant-General McNair was also killed in an (American) air raid during the Normandy campaign. The difference is statistically significant. In terms of size of army and duration of hostilities, the German army's experience was ten times greater, but the mortality difference was about 150-fold. In 1944–45 the US army knew that the German army was superior to itself in terms of combat performance.[13] Relative casualty rates, and other operational research data, support this.[14] Mortality rates among generals suggests a very different style of command, which may be a significant factor. Some German generals may have been killed in the encirclement battles of the latter stages of the war on the Eastern Front, but the numbers involved are not nearly enough to account for the discrepancy.

It seems that Britain has not been good at producing senior wartime combat commanders from its peacetime army. By comparison the German army seems to have been more successful, and the evidence points towards a very different command style. Do similar phenomena exist at lower levels? Individual combat experts can be identified in several disciplines. Their performance appears to be vastly superior to the average for their army, navy or air force in a given conflict. Is this phenomenon relevant, and how does it contribute to operational effectiveness? We shall look beyond land warfare in two cases – fighter-pilots and submarine captains – where useful comparisons can be drawn.

Much has been written about snipers and their apparently legendary effectiveness. Books on the subject tend to verge on idolatry and 'military pornography'. There are probably good anthropological reasons for this: the heroic image of the lone killer; the psychological impact of sniping; and propaganda. The duel between 'Noble Sniper' Zaitsev and Major Koenig in Stalingrad, made famous in the film *Enemy at the Gates*, probably did not take place.[15] However, it definitely did appear in Soviet propaganda, and (like much about the Battle of Stalingrad, or Kursk for that matter) seems to have been widely believed in the West.

Were snipers effective? Apparently so. German records from the Eastern Front seem to be well documented and reliable. The top-scoring Wehrmacht sniper was Matthias Hetzenauer of the Tyrol, with 345 certified hits. He was followed by Sepp Allenberger of Salzburg, with 257 hits. Helmut Wirnsberger of Styria recorded 64 before being wounded, and then served as an instructor with the rank of captain.[16]

Data for tank kills is harder to come across, although the Wehrmacht's experience seems to have been reasonably well documented in this case as well. The record seems to have been held by Michael Wittman, who died commanding a Tiger company in 1944 but had fought (largely on the Eastern Front) since 1939. He was ascribed 138 kills before his death near Cintheaux in Normandy.[17] No Westerner even approached this number. The reasons for the relatively smaller Allied success must include their having generally poorer tanks.

Similar statistics can be assembled for expert fighter-pilots. Glamour and prestige has accompanied them since the earliest days of aerial combat. A fighter ace is a pilot with five or more confirmed kills to his name. The top scorer of all time is Erich Hartmann of the Luftwaffe with 352. There were 104 other Germans who exceeded 100 kills.[18] No Western pilot reached 100. A significant proportion of German kills were Russian, but even where German pilots fought Westerners exclusively (such as Hans-Joachim Marseille with 158 kills) or predominantly (for example Adolf Galland with 104) they consistently outscored their opponents.[19]

Pilot numbers are so large that reliable statistics can be extracted. Aces are over-represented eight-fold among the kills. In the Western Desert the top 15 Luftwaffe pilots (from a total of over 200) accounted for 44 per cent of all enemy aircraft destroyed; an average of over 44 aircraft shot down for each expert pilot.[20]

A study of submarine captains shows a similar pattern. US submarines sank 1,150 Japanese merchant ships in the Second World War, but the top-scoring 25 boats sank an average of about 18 ships each.[21] That is, 39 per cent of Japanese merchantmen were sunk by about 8.5 per cent of the boats that served in the Pacific.[22] The top 25 boats were 6.77 times more effective than the rest. German U-boats showed a similar pattern. The top five boats in the First World War sank an average of 151 ships.[23] In the Second World War the top ten boats, which represent 0.87 per cent of the submarines, sank 7.7 per cent of the ships. They were about 8.85 times more effective than the rest.[24]

This effect is most marked when the rest are considered. The vast majority of fighter-pilots shot down none or perhaps one aircraft. Similarly the great majority of US submarines sank one ship, if any. There is therefore a hugely skewed distribution of kills among a few individual submariners and fighter-pilots. The same appears to apply to tank commanders and snipers.

From the standpoint of operational effectiveness, three questions should be asked. What produces these experts? Are they actually effective? Is the phenomenon worth fostering in order to improve military effectiveness?

First, a note of caution. We can be reasonably sure of many of these 'scores', but not certain. We must also question what they actually imply. The quality of air combat recording was discussed earlier. German sniper hits had to be certified by another observer. The number of certified hits was certainly smaller than the actual score.[25] For claims of submarine sinkings, a former U-boat commander found that, on the basis of over 20 years of comparing claims with convoy and escort vessels' reports, captains rarely invented claims. However, they sometimes had an over-optimistic evaluation.[26]

As to what they imply, a sniper 'hit' is just that; the sniper cannot normally know whether the target is dead or wounded. Sinkings are sinkings, whereas some 'aircraft shot down' are repairable. The available data did not allow the discrimination between submarine captain and crew. For example, an expert U-boat commander named von Montigny commanded five different boats in the First World War. However, one of those boats sank five ships under two different captains, another sank 22 under four, a third 49 under four, the fourth 11 under five, but the fifth 26 under 16 different skippers![27] It is possible to separate actual scores achieved by given captains, but the figures do not seem easily available.

What produces these tank, sniper, submarine or fighter experts? The answer can be considered under two headings: the individual and the organizational. There is a respectable body of psychology known as 'expert theory'.[28] Broadly, an individual can become expert if three conditions apply: the absence of disqualification; the ability to learn; and the tendency to learn. Skill of itself is not a factor. 'The absence of disqualification' appears curiously negative. It implies any practical exclusion from doing well. Poor eyesight will probably exclude many from being snipers or fighter-pilots, because even with glasses their peripheral vision is impaired, and spectacles do steam up at critical moments. In the Second World War, Western tank commanders were systematically penalized because even if they hit the target, there was no guarantee that a hit meant a disabling kill against a Tiger or a Panther. Failure to disable might mean failure to survive. Would-be submariners who are chronically seasick will probably not do too well.

The ability to learn reflects the opportunity to gain experience: a fighter-pilot will not become proficient if he never meets an enemy aircraft, nor a submariner a merchantman. Some experts were fortunate to have 'target-rich' but benign environments when they first plied their trade. All three German snipers described earlier were woodsmen from the Alps. They had probably learnt to handle a rifle at an early age. Yet opportunity is not the whole answer. Although Galland first flew in combat in the Spanish Civil War, Hartmann started only in October 1942 and Marseille flew in combat for a total of just 28 months.[29] The three Wehrmacht snipers listed above only started sniping in 1942 or 43.

The tendency to learn is probably more critical. It implies a tendency to reflect on experience and learn from it, to maximize the benefit of experience. This characteristic enables one to *develop* skill. Hence skill is an effect, not a cause. For example, snipers tend to be very single-minded people, and this does not just focus their attention on the target at the point of engagement. It also focuses them on the business of sniping generally, to an almost fanatical degree. *Almost* fanatical; the virtues most prized in a Wehrmacht sniper were clear thinking, calmness, judgement and patience.[30]

I once worked with one of the very few Master Sniper instructors in the British Infantry.[31] His mental powers of detachment were almost as remarkable as his shooting, which was phenomenal. He did not have a particularly high IQ, but he could be curiously engrossed at times. His ability to concentrate on one activity was remarkable. He could also coach others to shoot to very high standards with relatively little opportunity to train. That reflects an acute awareness of the critical details required to achieve a high level of skill.

The three factors of absence of disqualification, ability to learn and tendency to learn can produce skill levels greatly above the average. The statistics available from large samples – fighter-pilots and submarine captains – support this. That suggests that the same is true of other individual skills such as commanding a tank or sniping. There is some hard evidence, from both field trials and war records. In British field trials conducted in the 1980s, the top 20 per cent of tanks fired roughly half of all rounds fired. In the Second World War, the top 10 per cent of British anti-tank guns typically knocked out ten times as many enemy tanks as the rest.[32]

Organizational factors also play a part, be they sociological or cultural. Western air forces in the Second World War adopted a '50-mission' policy with aircrew, in the seemingly reasonable belief that without some credible end to tours of operational flying duty, the aircrew would break down psychologically. Thus Western fighter-pilots were in practice limited in their opportunities to score, simply because they were 'rested' at prescribed intervals. This gave their opponents both more opportunity to score in absolute terms and more opportunity to learn from experience.

This begs the question why Germans didn't have the equivalent of a '50-mission' policy. Hartmann flew almost continually for 2½ years and was still shooting down aircraft on the last day of the war. The same was true for Hans-Ulrich Rudel, the dive-bomber expert, and several other German pilots. They didn't crack up. The Germans were well aware of the possibility of 'combat exhaustion'. They sent pilots to a sanatorium at Bad Wiese to recuperate.[33] After the war surprising numbers of them appeared to be likeable, well-adjusted and alert.[34] It appears that instead of a 50-mission policy, German COs were educated as to the symptoms of combat exhaustion, and made aware of the appropriate remedies. Those remedies included extensive leave or a tour of duty as an instructor. Thus they were able to manage their aircrew's well-being and get the best from them. Since 'the best available' from the 'best of the few' was very good indeed – and far better than their Allied peers – one is left with the impression that the 50-mission policy actually reduced operational effectiveness.

The second question is whether such experts have any significant operational impact. Hit, kill or sinking numbers represent outputs rather than outcomes. Effectiveness is not the same as efficiency. Can we actually say anything more than that the statistical distribution of kills among killers is highly skewed?

It appears we can, for at least two reasons: their tactical impact and their impact on morale. For example, a Wehrmacht infantry battalion had up to 22 snipers. On the Eastern Front, four to six snipers would typically form a rearguard in the withdrawal. Relatively few hits were normally sufficient to break the momentum of the Russian infantry attack or advance sufficiently for the German battalion to break contact.[35] A few experts therefore enabled whole units to withdraw with little risk. In the attack snipers engaged enemy commanders and support weapon crews. The effect of this on the physical cohesion of a defence can be imagined; it is further heightened by psychological impact. As well as being frightening, sniping is hugely distracting to commanders and degrades low-level command and control.[36]

The same appears true of tank commanders. Wittman's counterattack at Villers-Bocage in June 1944 is now legendary.[37] In essence, one tank defeated an armoured division. The 4th County of London Yeomanry suffered 24 tanks destroyed in minutes; the 7th Armoured Division's attack stalled; and a threat to 1st SS Panzer Corps' rear was averted. The key lies in the latter: it is not the number of tanks – a bad day's fighting for an armoured regiment in Normandy – so much as the overall tactical impact, which had operational and even strategic consequences. Wittman effectively managed to inflict shock on an armoured division! Despite the quality of scorekeeping, it is difficult to observe similar tactical impact for fighter-pilots or submarine captains. There is no obvious case of an 'ace' of either discipline having such tangible tactical effect.

The other aspect of effectiveness, and a critical one, is the impact on morale. In practice much of the effect of snipers is psychological: the reasonable man will not get up and engage the enemy if he believes there is a good chance of being hit by a sniper. The average rifleman can have no real idea whether there is such a chance or not. The evidence of someone near him being hit is more than sufficient. That can bring whole battalions to a halt, and it takes a lot to get them going again.[38] Similarly it would only take one or two brushes with a Tiger for a Western tank crew to develop a very cautious and healthy respect for them.

The case of fighter-pilot Hans-Joachim Marseille is a further, useful, and possibly contradictory, example. He arrived in North Africa in the spring of 1941 with a tally of seven kills, and scored 151 further kills in the next 18 months. He was much fêted by German propaganda as 'The Star over the Desert'. Certainly 158 kills is a high total, but it averages only eight or nine a month. Although that is impressive, we should note that the RAF lost 31 Hurricanes over Greece in one month in 1942.[39] It had lost 136 aircraft *in one week* in September 1940 during the Battle of Britain, and kept flying. There was only one Marseille, who couldn't be everywhere, whilst there were 47 RAF and Commonwealth fighter squadrons in the Middle East in 1943.[40] There appears to be no strong evidence of Marseille's (nor any other individual fighter-pilot's) effect on enemy morale in the Second World War.

The propaganda aspect illustrates the beneficial effect on friendly morale: the 'Noble Sniper Zaitsev' episode, Marseille, and even the habit of parading Aces around aircraft factories, all reveal a simple, clear and easily exploited message: 'look how many bad guys we can kill!' The relatively low scores achieved by the vast majority of their peers may not have been apparent either to the propagandists nor their audience, and is probably irrelevant here.

The third question is whether extreme combat expertise is worth promoting. Ostensibly, since it has real operational benefit in terms of outcome (effectiveness) and not just output (efficiency), the answer is 'yes'. We should, however, also consider inputs and emergent properties. Firstly, what do you have to do to produce the few experts, over and above what you do with the rest? If we consider snipers as expert infantrymen, the British Army considers four to six weeks' specialist training as being sufficient to produce technical proficiency. However, skills fade very quickly. When I took an informal poll in 1991 I did not find any regular British infantry battalion which maintained a body of snipers who requalified annually as required. The standards seem to be very difficult to sustain in practice in peacetime. Not least, the opportunity for a sniper to learn from experience in peacetime is very limited.

The real cost in this case seems to be to have some soldiers trained when the battalion enters conflict, and then use them extensively when the situation permits. This is arguably a low cost. Similarly with fighter-pilots (and tank-busters such as Hans-Ulrich Rudel): the real cost seems to be a permissive man-management regime that allows those who *can* learn rapidly from their experience to continue to do so. Hartmann does *not* seem to have been deeply traumatized by his experiences. He went on to be a general in the new Luftwaffe after 1955.

Thus there does appear to be real value in promoting individual excellence. The real costs are human, and arguably low. Some of the benefits are also human, in terms of psychological and morale effect, both on the enemy and one's own side. One cannot really predict who will be an expert, although some psychological characteristics may be informative. The Master Sniper I knew had some fairly obvious character traits, but never had the opportunity to fire his rifle in anger in 22 years' service. In war, one could perhaps select such individuals and post them where they could gain experience, at cost to the units losing them.

The whole subject has yet to be explored systematically, and the evidence is scant. The emergence of experts seems to be very largely an unforeseen consequence of personnel management policies. The right policies produce tactics, organizations and processes by which a few people kill a lot of the right people at the right time and place, and have significant battlefield outcome. Clearly, further study is needed.

We now turn from the individual to the sociology and culture of an army. What impact do these factors have on the way commanders think, and how does that influence the effectiveness of the army in which they serve? As an example, we shall examine the British Army, looking at some of the organizational factors that underpin the way that it has performed and changed over the years.

Armies do introduce new technology in peacetime, and technology normally brings improved effectiveness. The British Army of the 1970s had tanks with ranging MGs; recoilless anti-tank guns; and passive night sights. The Army of the

1990s had laser range finders, ballistic computers, ATGW, and thermal imaging sights. Conversely, we have suggested that it forgot many hard-won lessons from the Second World War in the long years of subsequent peace.

Combat is essentially human; we should look for the reasons in the gross, organizational, human aspects of an army. Those reasons seems to lie in the lives and careers of the people in it. Deep, hard-learnt operational effectiveness built on collective performance and cohesion appear to be ephemeral and almost fragile. Veteran formations and units are curious ephemera. The poor performance of the US army in the first months of the Korean War illustrates how the effectiveness of US units had been seriously eroded between 1945 and 1950.[41] Christopher Duffy similarly highlighted the passing of the knowledge and experience of the Seven Years' War by the time of the Napoleonic Wars.[42]

As a slightly more prosaic example, consider the rifle company which I commanded from 1994 to 1996. I assumed command of the company just before Christmas 1994. Once warned for operations in Northern Ireland, the company was brought up to strength with a full complement of officers and soldiers. Many of the privates were straight out of training. The company trained extensively, deployed from England to south Armagh in March 1995, and returned with almost exactly the same personnel with which it deployed. By late October 1995 the company had returned to England and taken disembarkation leave. At that point it was a cohesive and coherent company, with lots of shared experience, which worked well as a team.

I left in December 1996. At that point none of the officers and only two sergeants, three corporals, three lance-corporals and less than 20 of the 50 or so privates that had served in the company in south Armagh were still in it. The privates had all been the most junior at the time. On most of the dozen or so occasions that the company trained together after the Northern Ireland tour, I gave orders to a different set of platoon commanders. The three platoons were commanded by ten different people during those two years. The 'surviving' sergeants and corporals were the backbone of the company. They could remember how things were done in south Armagh and what mattered, as well as what didn't. They were still there in early 1997, but by then a new company commander and sergeant-major were in post. Thus after two years the company had in practice lost all the collective performance and social cohesion it had gained in Northern Ireland.

What remained, of course, was a lot of individual experience and skill. That would be remembered, and passed on for many years to come. It trained for different roles in different theatres; no doubt to the same or higher standards. But the company would never perform again *in the same way*. If a permanent ceasefire had occurred in 1996 and no further British Infantry battalions had been sent to Northern Ireland, the subsequent history of that company suggests that the

British Army would have lost all of its first-hand collective performance within two years. This suggests why veteran units and sub-units are ephemeral.

It does not necessarily mean that the army as a whole loses its edge so quickly. The original company Second-in-Command was passing on his experience at a training establishment within weeks of disembarkation, as was one of the corporals. That benefit persists for many years. The example also highlights the need for collective training. In early 1997 the company contained many capable officers and NCOs, but it had lost the particular set of collective skills which it had in Northern Ireland two years previously. To bring that back for a subsequent tour required two or three months' training.

British soldiers normally serve for no more than 22 years. A corporal in 1995 might be an instructor as a senior NCO or warrant officer about 10 years later. He might leave the army in 2013 or so. A young captain of the same vintage might serve until 2027. By comparison, Colonel 'Tod' Sweeney of the Oxfordshire and Buckinghamshire Light Infantry won a Military Cross (MC) as a platoon commander at Pegasus Bridge in Normandy on D-Day. He was deputy commandant of the School of Infantry as a colonel in 1971–73, and left the army in 1974.[43]

Other examples include Major-General Matt Abraham, who joined the army in 1939. He was Commander, Royal Armoured Corps in 1st British Corps in 1965–67 and retired in August 1973.[44] Major-General James Lunt was commissioned in 1937; he was the Vice Adjutant-General from 1971–73 and retired in August 1973.[45] Major-General Colin Purdon commanded the North-West district in 1972–74.[46] He had won an MC at Saint Nazaire in 1942. (Incidentally, he and his staff car driver won the district light machine gun trophy in 1973 – a rare feat for a major-general!)[47]

A very few generals serve through to the age of 65 – a total of 47 years' service. Thus a general who joined in 1944–45 could have served until 1991. The vast majority of officers retire at or before the age of 55. Almost all veterans of the Second World War would therefore have left the army by 1982. By 1970 few battalions would have any officers or soldiers with wartime experience.[48] By 1980 there were probably only a small handful of officers in the British Army who had served in the Second World War. (This technique of considering the length of lives or careers is known in sociology as 'longitudinal studies'.)

The generation who served in the British Army in the decades immediately after 1945 had considerable experience of other conflicts, such as Palestine, Korea, Borneo, Malaya, Cyprus, Aden and Northern Ireland. Only Korea included extensive warfighting operations, and only Northern Ireland involved the vast majority of the army. Thus although the army (and Royal Marines) which found itself responding to the Argentine occupation of the Falklands was very soldierly, it had very little direct experience of general war. That was enough: it knew enough of general war to beat the Argentinians. As a newly joined subaltern in the spring of 1982, I vividly remember the key phrase used to describe the lessons from the Falklands: 'No new lessons, just old lessons relearnt.' That now seems completely unsurprising.

Individual experience generally persists at company level for up to ten years: the Company 2ic of 1995 commanded a company himself in 2003–4. Veterans might command battalions up to 20 years after a conflict, and formations 30 years. Their commands would reflect their experience: a company in the Korean War might have been commanded by a veteran of the Second World War. He might have commanded a company or even a battalion in 1939–45. By the 1970s only the somewhat indirect effect of a few senior individuals (such as Colonel Sweeney at the School of Infantry, or General Purdon in north-west district) would be felt. There would be collective memory, but little of the hard-learnt collective performance.

We might call this process a '30-year rule'. By analogy, the generals in the US army of the 1990s were the company-level officers who fought in Vietnam in the 1960s and into the 1970s, General Norman Schwarzkopf being an example.[49] The generals of the Vietnam era were commissioned in the late 1930s and early 1940s. Of the 456 members of the West Point class which graduated in 1939, 224 became (full) colonels and 72 became generals. One became NATO's Supreme Allied Commander, Europe; one commanded I Field Force in Vietnam; and at least one served to 1978, 39 years later.[50]

The experience of the armies of the Middle East makes an interesting comparison. They fought major campaigns in 1948, 1956, 1967 and 1973. Thus by 1973 they would have veteran company, battalion *and* formation commanders. Major-General Ariel Sharon, latterly the prime minister of Israel, commanded a division in both 1967 and 1973. By 1973 the Arab and Israeli armies were probably as experienced throughout the chain of command as any since the end of the Napoleonic Wars.

There is therefore some form of generational effect. The career of Field Marshal Sir Nigel Bagnall, considered to be the father of British Military doctrine, is an example. He joined the army in 1944 and was commissioned in February 1946. Bagnall commanded the 4th Armoured Division in 1975–77.[51] Thus the man considered to be 'the father of British military doctrine' was one of the first divisional commanders not to have held a commission during the Second World War. An officer even one year older could have seen action in that war, and every year older would have brought more wartime experience.

Bagnall's concrete contribution to army organization was to insist on significant reserves being available at divisional, corps and then army group level. He was the officer most responsible for moving the British Army away from the small, Second World War-pattern divisions. Thus the first major thinker of the postwar generation was the officer who broke the war-learnt pattern of small, agile divisions. This may seem far-fetched, but there are intriguing parallels. The first General Secretary of the Communist Party of the Soviet Union to have a university degree (since Lenin) who was not a survivor of the purges of 1936–39 was Mikhail Gorbachev, the architect of Glasnost and Perestroika. His predecessors – Brezhnev, Khruschev and Andropov – were all young apparatchiks in 1936.

More prosaically, my older brother joined the TA in 1972. At that time infantry tactics were described by highly experiential epithets such as 'KISS' ('Keep it Simple, Stupid'), 'Two up and bags of smoke', and similar. When I joined the TA as an officer cadet in 1978, such epithets were derided as simplistic – they probably were to some extent in 1972 – and were falling from use. They had largely disappeared when I attended the platoon commander's course in 1983. Many of the officers who had commanded platoons in the Falklands conflict attended the same course. It can be suggested that the generation who fought in the Second World War, and those they taught directly, believed in simple but effective drills well executed. Their successors did not really understand that.

Earlier we considered the very poor operational record of the commanders of the BEF of 1940. We suggested that similar failings had occurred at the outbreak of the Boer War and possibly also in 1914. It appears that many of those whom the British Army promoted in peacetime during the twentieth century were found wanting on the outbreak of war. Promotion to high command in peacetime very much reflects the values of existing senior commanders, themselves largely the products of a peacetime promotion system. To that extent it reflects deeply held values, and has an obvious impact on operational effectiveness in war.

Roughly two-thirds of those who commanded formations in the BEF of 1940 were either sacked, retired immediately, or were never given another formation to command in the field. There is evidence of a similar process amongst more junior commanders. In the HQ of 21st Army Group, still in Britain just after D-Day, 'they talked *sotto voce* for a while about the crop of adverse reports which had come back from France, against hitherto successful battalion commanders and brigadiers who had lost their head when the guns begin to fire'.[52] Furthermore, the Army List for 1946 shows that most of the British corps and army commanders of 1945, and the very senior staff, were at most brigadiers in 1939.

In 21st Army Group of 1944, only the Commander in Chief (Montgomery) had commanded a division or more in the BEF. Slim was a brigade commander in the Indian army at the outbreak of war. At least nine of those who subsequently reached the rank of lieutenant-general or higher had been lieutenant-colonels, colonels or brigadiers in 1939: Generals Dempsey, Leese, Horrocks, Crocker, Gale, Browning, Templar, Slim and Harding. They had been promoted at least twice during the war. That is in fairly stark contrast with the 'best available' for 1940.

Almost the whole raft of British commanders from divisional command upwards in 1939 had been found wanting.[53] The exceptions were notable: Alanbrooke, Alexander, Montgomery and very few others.

Applying the '30-year experience rule' formulated above, the formation commanders of the BEF were typically commissioned in the mid 1900s. They would have been subalterns or captains in 1914. Given the rapid expansion of the army and high officer casualties in the First World War, they would have been promoted rapidly and gained considerable wartime experience. Most of them were highly decorated. Between the wars the British Army suffered from force reductions, underfunding and the possibility of being 'retired on half-pay' if not promoted. This environment seems to have bred conservatism and rewarded those who appeared to correspond to the prevailing stereotype.[54]

There is something in stereotypes. It may be unfair to tar everyone with the same brush, but it is probably safe to make general comments about groups of individuals as a whole. For example, the junior officers of the British Army in the 1920s and 30s appear as somewhat anti-intellectual. Whilst a cadet at Sandhurst in the 1930s, the future Major-General James Lunt gained full marks in one exam. Whilst enjoying a round of golf, the Commandant said to his father, 'I don't think we should have too many brilliant officers in the army . . . What we want is good young men who can ride well, get on with their soldiers and behave properly.'[55] At the highest level there is a stereotype of the charismatic, hugely personable leader as Commander-in-Chief. This stereotype is interesting because it stands almost no scrutiny. It may be true of Slim (of whom more below), but not of Montgomery,[56] nor Wellington,[57] probably not Manstein,[58] nor Rommel,[59] Guderian,[60] or Zhukov.[61]

Perhaps the acomplishment of interwar success was Lord Gort, Commander-in-Chief (C.-in-C.) of the BEF in 1940. He was a highly decorated, charismatic and aristocratic young major-general in 1937. Field Marshal Sir Cyril Deverell's sudden sacking saw Gort wished on the army as Chief of Imperial General Staff (CIGS) by Hore-Belisha, the Secretary of State for War.[62] Gort become C.-in-C. of the BEF on mobilization. The Kirke Report had recommended that CIGS should *not* become the C.-in-C. on mobilization.[63] Gort appears to have been sent to France because he was ill-suited to military politics.[64] Nevertheless, he was the epitome of the system: young, highly decorated, charismatic, promoted through and entirely within the system. He was only 51 when appointed CIGS. Gort was effectively removed from public sight after the Fall of France. As C.-in-C. of the BEF, he 'fussed over details and things of comparatively little consequence' and had a 'constant preoccupation with things of small detail'.[65]

Meanwhile, the 47-year-old Bill Slim was promoted to lieutenant-colonel in 1938, perhaps at the last possible opportunity. Slim had not been to Sandhurst; he had gained his commission 'through the back door'; and he came from a modest background.[66] The outbreak of the Second World War saw him commanding a brigade in East Africa. Within four years he was commanding the Fourteenth Army in Burma. His career continued to prosper after the war; he succeeded Montgomery as CIGS and was promoted to Field Marshal at the age of 57.[67] Slim was obviously *not* the product of a stable hierarchy in peacetime. His rise to fame came entirely during wartime. He was arguably one of the greatest British generals of the twentieth century. The contrast with Gort could not be more marked.

Two theses can be advanced to explain the failure of most of the commanders of the BEF: unpreparedness and unsuitability. The BEF was poorly prepared in various respects. The army received low priority for resources in the 1930s. Its standard of training appears to have been patchy, if not poor.[68] There appears to have been little training of HQs, so formation commanders may have been poorly prepared for their responsibilities.

The resources required to train commanders and staff were relatively modest compared with those required for re-equipment. They were typically a few maps, telephones, intellectual effort and time. The German army did commit such resources: the crossing of the Meuse at Sedan had been studied well in advance (and even rehearsed on the Mosel). The Kirke Report was not published until 1932. Conversely, the German CGS, General Hans von Seeckt, had ordered several studies into the reasons for the failure of the German army in 1914–18 in 1920.[69] The difference appears to be not so much one of unpreparedness, as lack of disposition to prepare.

The commanders of the BEF were highly experienced veterans of the First World War. The same is true of the commanders of the Wehrmacht. But experience can be misleading if one does not reflect on its relevance. Fuller definitely did; likewise Rommel, Guderian and Manstein. The evidence that the commanders of the BEF of 1940 had done so is not strong, and we have seen that a tendency to learn, to prepare oneself, is a characteristic of an expert. It is illustrative that in 1940 Fuller had retired, while Rommel, Guderian and particularly von Manstein were in key positions in the Wehrmacht.

This suggests not that the BEF was unprepared, but that its commanders showed little disposition to prepare themselves. That finding supports the broad thrust of this book. It shows the benefit of a deep and broad exploration of the conceptual component of fighting power. It seems that the interwar British Army was not disposed, for human reasons, to make such an exploration.

I was a junior officer in BAOR in the 1980s, and noticed that many senior officers had a reputation for being bad-tempered bullies and sackers. A former officer who was commissioned in 1959 and retired as a colonel in 1994 recalls a generation of formation commanders in the late 1970s and early 1980s who 'felt they had to rule by fear'.[70] He pointed out that many of their immediate seniors had wartime experience, and were often charming. He cited General Sir John Hackett as an example. Conversely, their immediate juniors were often better educated. Their own education may have been disrupted by the war. A typical authoritarian response would be to feel challenged, and respond through sacking and bullying.

Why did this happen? Do armies inevitably attract bullies, or is there more to it? We might look first at the First World War. Haig preferred young, dynamic, forceful brigade commanders; such officers rapidly gained a reputation as 'thrusters'.[71] They were typically majors or perhaps captains in 1914. This was the generation that produced Lord Gort: he was 28 in 1914 and commanded 3rd Guards Brigade briefly during 1918. To some extent the same thing happened in the Second World War. Carver was commissioned in 1935 and commanded 4th Armoured Brigade in 1944–45 aged 29.[72] Carver gained a deserved reputation as a sacker. He sacked four COs, from a total of only five units in his brigade.[73] He was quite a role model in the postwar period, becoming Chief of the Defence Staff in 1973.

The postwar army suffered a series of reductions in 1948, 1958 and 1969. It was full of senior officers with distinguished war records but, as we shall see, the path to high rank lay mostly in the command of armoured formations in BAOR. That was a peculiarly stable military environment in which the 'thrusters', who were quite prepared to bully and sack people, could easily rise to the top. People such as such as Carver would have served as role models. One senior officer, at the time a captain, who was sent on an operational detachment in 1979, remembered being advised to 'sack someone early' as an example to the rest.[74] The promotion rules of the time, which allowed officers to jump straight from lieutenant-colonel to brigadier, probably exacerbated the phenomenon of the 'young thruster'. A 32-year-old captain could become a 42-year-old brigadier chiefly by making the right sort of impression in his first tour of duty as a major after Staff College.[75]

It would be easy to gain the wrong impression of what such officers were like. To fit the stereotype they had to appear to be gentlemen. They rarely if ever raised their voices, and very rarely swore. They could bully mercilessly without doing so. One very subtle characteristic was the ability never actually to tell someone what to do. This was probably subconscious and had three aspects. One was that giving direct orders might be seen as ungentlemanly. The second was that if one didn't actually say what was wanted, the chances of a subordinate getting it right were impossibly low (an advantage to an inveterate bully). The third aspect was that such men get along by bullying. They may not have known, nor particularly cared what they wanted the subordinate to do. The key thing was to rule through uncertainty and fear.

On occasion the guard would slip. Not that the bully would raise his voice or swear; he could lose his temper most effectively without doing so. On one occasion a highly socialized bully was contradicted in public on a point of fact by an officer considerably his junior. In response he delivered a monologue which lasted over 12 minutes. The ability to do so without preparation was quite remarkable. It suggests 'driven' behaviour, cognitive dissonance, or some other moderately unusual psychological phenomenon. Research into his background revealed several unexpected examples of vicious and vindictive behaviour in the ranks of lieutenant-colonel to major-general. These instances are drawn from the British Army but I know from first-hand experience that authoritarian bullies can also be found elsewhere: in the US army, for example. They thrive in the long shadows of a major war. Regrettably, they themselves can also cast a long shadow.

9

Regulators and Ratcatchers

We have seen that undesirable personality types might succeed, and indeed may have succeeded, in an army in peacetime. The consequences can clearly be catastrophic. But why do such people prosper in peacetime? Could they be identified? The answers to such important questions are, unfortunately, not categorical. However, some aspects of psychology do provide some partial answers.

There is no single, generally accepted theory of personality. Freud considered personality to be innate, whilst modern theory is more descriptive. The more easily measurable aspects of personality describe either trait or temperament. A trait is, effectively, a single measurable parameter of behaviour, and there are several different taxonomies of traits. Temperaments are observable groups of traits which appear to be connected.[1] There are also other theories such as the behaviouralist approach. Personality theory seems to be relatively undeveloped in terms of the history of a science.[2] There are several overlapping descriptive schemes. There appears to be little understanding of causality, so existing theory is not particularly predictive. The most successful aspects of prediction are long-term and broad in scope.[3] Personality theory cannot yet predict in detail how given individuals will behave in given situations.

In the early 1950s the American psychologist Isabel Myers and her mother Kathryn Briggs developed a system of assessing personality which is now widely used and respected. Their Myers–Briggs Type Indicator (MBTI) scores subjects on four scales as either 'Extrovert' or 'Introvert', 'Sensor' or 'iNtuitive' (*sic*), 'Feeler' or 'Thinker', 'Judger' or 'Perceiver'; hence E or I, S or N, F or T, J or P for short.[4] The one-word titles are not particularly useful. For example, an Introvert draws his energy from within, rather than from interpersonal relationships. Introverts are therefore not necessarily shy. They may or may not be, depending on how well socialized they are. Many will be: their temperament gives them little predisposition to seek interpersonal relationships in youth and early adulthood.

MBTI results in a four-letter description of each personality type, such as INTJ or ESFP. There are useful short character sketches for each of the 16 combinations. There are also one-word titles, which are again also not particularly useful as descriptions. They are, however, a useful shorthand.

Character types are not equally represented in the population. Most people are extroverts, hence Es outnumber Is. Men are equally divided between Fs and Ts, whereas more women are Fs than Ts. A list of the 16 types, with their prevalence in the population, a short title and the sorts of professions they tend to join is given in Table 4.[5]

Table 4 MBTI

Type	Short Title	Proportion of the Population	Typical Profession or Occupation
ESFP	Performer	> 10 per cent	Singer, comedian, PR, politician.
ISFP	Composer	< 10 per cent	Composer, artist, chef, fashion designer.
ESTP	Promoter	≈ 10 per cent	Salesman, property developer, show business producer.
ISTP	Crafter	≈ 10 per cent	Surgeon, racing driver, skilled craftsman, pilot.
ESFJ	Provider	> 10 per cent	Caterer, financier, retail tailor.
ISFJ	Protector	< 10 per cent	Underwriter, shepherd, caretaker.
ESTJ	Supervisor	> 10 per cent	Foreman, policeman, referee.
ISTJ	Inspector	< 10 per cent	Auditor, judge, inspector, surveyor.
ENFJ	Teacher	≈ 2 per cent	Teacher, journalist, clergy, therapist.
INFJ	Counsellor	≈ 1 per cent	GP, tutor, psychiatric nurse.
ENFP	Champion	2–3 per cent	Novelist, orator, playwright.
INFP	Healer	≈ 1 per cent	Social worker, librarian, missionary.
ENTJ	Field Marshal	< 2 per cent	Senior military officer or chief executive.
INTJ	Mastermind	≈ 1 per cent	Planner, coordinator.
ENTP	Inventor	< 2 per cent	Project manager, engineer.
INTP	Architect	≈ 1 per cent	Mathematician, scientist, logician.

Temperament and IQ are not related. Hence, for example, there are probably about 600,000 INTPs in Britain, of whom only those with an IQ of about 120 or higher – the top 10 per cent – will be capable of qualifying as mathematicians, scientists or logicians.

MBTI appears to be a valid and useful description of personality. David

Keirsey developed it into a more usable tool. He realized that four groups of MBTI types had strong similarities. He called those groups 'Temperaments', depending on whether they contained the S and P (hence SP), SJ, NF or NT characteristics.[6] He assigned one-word short titles to those groups, which should also be used with caution:

> **SPs: the Artisans**. SPs are all broadly creative and artistic. The key issues which drive their temperament are performance and skill: be it as a skilled craftsman; an artist; a surgeon; or a jet pilot. About 40 per cent of the population are SPs.

> **SJs: the Guardians**. SJs are focused on the (human) institution: running, maintaining and protecting it. In consequence much of their thinking is rule-based. Rules, maintaining and applying them, are important of themselves to Guardians. Status tends to be important to them. Another 40 per cent of the population are SJs.

> **NFs: the Idealists**. NFs are friendly, empathic 'people people'. Personal relationships and maintaining them tend to drive Idealists. NFs form 10 per cent or so of the population.

> **NTs: the Rationals**. NTs are logical, rational and analytic. The rationale, the method, the plan and the understanding tend to dominate the behaviour of Rationals. They tend to appear logical and unemotional. NTs form the remaining 10 per cent of the population.

These brief summaries appear trite and unconvincing here. Not least because the population does not fall neatly into four quarters: SP Artisans and SJ Guardians together form 80 per cent of the population, and some individual types only form about 1 per cent of the population. Furthermore, temperament is not personality. In practise it takes a skilled and experienced observer to observe temperament or type directly, rather than by scientific assessment.[7]

The way in which people behave within a human organization will depend on their temperament. For an SP ('Craftsman'), the organization will provide systemic opportunities to practise or display their skills: surgeons in hospitals, fighter-pilots in air forces. For SJs ('Guardians') organizations are things to be run, organized, protected and maintained. For NFs ('Idealists'), they are places where relationships happen. To NT ('Rationals'), they mean relatively little, except perhaps as objects of study. However, since organizations are human and not overtly rational, they do not necessarily attract NTs' attention.

Temperament is strongly correlated with choice of job. A high proportion of teachers are NFs or SJs. About 25 per cent of nurses are NFs; more than twice as many as in the population as a whole. It would not be surprising if army officers were predominantly SJs. That might be a good thing, as SJs are predisposed to protect, maintain and nurture human organizations, and be adept at creating and obeying the rules needed to make them function.

Professor Norman Dixon identified two relatively distinct types amongst generals, which he described as authoritarians and autocrats. Dixon identified many cases of apparent military incompetence as being caused by authoritarian commanders who had typically risen to senior rank in peacetime. Military historians tend to deride Dixon's work on grounds of historical inaccuracy.[8] Conversely, the psychologists with whom I have discussed his work agree that it is psychologically respectable. His history may not be particularly accurate, but his general findings are very insightful. Authoritarians are controlling, highly conformist, status-conscious, anti-intellectual and punitive of others who do not conform as they do. At a deeper level they are inhumane. They tend not to consider humans as people, but as objects whose importance is related to their status. Conversely, autocrats might also exercise tight control, but only when the situation demands it. They are forceful, driving and not particularly empathic. They differ from authoritarians in two critical areas. Firstly, they tend to be thoughtful and reflective. Secondly they demonstrate a deep humanity, even if only because they realize that the welfare of the people under them is vital to their effectiveness.[9]

The British naval historian Andrew Gordon's book *The Rules of the Game: Jutland and Naval High Command* identified a dichotomy within the characters of senior Royal Naval commanders in the late Victorian and Edwardian eras. Gordon discriminated between 'Ratcatchers' and 'Regulators'. The term 'ratcatcher' is a direct quote from Admiral Sir David Beattie, probably the greatest autocrat of Royal Navy in the early twentieth century. Gordon saw 'Regulators' as authoritarians, who generally shared a fairly closely defined social and professional background.[10] That background implies much about a fairly stable pre-war naval hierarchy. The details of that background are not too important. The important issue is that the Regulators were conforming to a particular set of social expectations.

The authoritarian personality was first identified by British psychologists immediately after the Second World War, by studying the personalities of extreme Nazis.[11] The landmark study into the subject was published in 1950. It developed a personality test to identify the type. The study was reviewed in 1996: both the type and the test remained valid.[12,13]

Fundamentally, authoritarians need to control their environment. They are very intolerant of uncertainty. They may appear charismatic, but not necessarily. Authoritarians will tend to thrive in highly organized hierarchies, such as armed forces in peacetime. They also want to be loved and do not brook argument. Dixon considered their need to be loved to reflect emotional pressures in childhood, possibly at the hands of socially insecure and therefore strict parents.[14] That need to be loved will be very deeply concealed and may be directed towards their military superiors, not least because authoritarians are highly status-conscious.

Authoritarians are often bullies. This has several effects. One is to damage the working environment within an HQ. Another is to discourage the bearing of bad news: they tend to 'shoot the messenger'. This obviously skews the effectiveness of command. Authoritarians also tend to sack people. Since they are driven by status and the need to progress, they may feel the need to make a visible impact on their commands during their tour of duty and do so by making rapid changes, such as sacking people.[15] Alternatively, authoritarians may be intolerant of people they perceive to be violating their rules, or merely their expectations. A senior officer who arrives at a poor or mediocre organization and sacks a number of subordinates may well be authoritarian. One who sidelines a few whilst getting the best out of the rest is more likely to be an autocrat.

Field Marshal Carver does not appear to have been authoritarian, despite his reputation as a sacker. In his mid and late twenties throughout the Second World War, he seems to have realized that war is too important to suffer fools or the incapable: 'You don't win wars by being kind. In war, ruthlessness comes out on top of kindness. You must be tolerant, but not tolerate idleness, stupidity etcetera.'[16] Additionally, he commanded a brigade at the age of 29. His immediate subordinates would typically have been older than him and served longer. Sacking unit commanders may have been necessary to impose his authority. Alternatively it may have been the reaction of a relatively young man who, with greater experience, might have done things differently. Nonetheless, he and his peers would have been role models for the postwar generation.

The generals of the 1980s would have joined the army just after the Second World War and spent much of their career in that stable, peacetime army. In practice, in order to command BAOR one had to have commanded 1st British Corps; hence an armoured division; hence an armoured brigade; hence an armoured, mechanized, artillery or engineer unit *in BAOR*. Command of BAOR usually led to the post of CGS. It is difficult to imagine a more stable, formalized hierarchy existing and promoting from within for more than a generation.

Combat is fundamentally uncertain, and authoritarians have difficulty in coping with uncertainty.[17] In war this may cause enormous stress, leading to a failure to cope and thus all-round failure. Alternatively, an authoritarian may be demonstrably poor at his job, so he will be sacked. He may take decisions too early in an attempt to reduce uncertainty.[18] Or, he may try to seek more and more information to try to reduce the uncertainty. This will result in bigger and bigger HQs. Such growth will be visible in peacetime, and may explain some of the enlargement of HQs since the Second World War.

Slim read extensively into his profession.[19] Therefore it is unlikely that he was an authoritarian. Authoritarians are markedly anti-intellectual: they dislike conceptual issues. This places Slim on a par with Wellington, Rommel and (perhaps surprisingly) Patton.[20] Patton's personality is interesting. He definitely displayed strong evidence of authoritarian behaviour. However, behaviour and personality are not the same.[21] The American writer Martin Blumenson's somewhat hagiographic biography reveals Patton as a hugely driven character, but that may be explained by well-documented learning difficulties as a child and youth. Overcoming those difficulties, which he clearly did, may explain authoritarian behaviour but a non-authoritarian personality. This may help us to understand Patton as a somewhat complex individual. It is also important to distinguish between a propensity for facts and intellectualism. SJs are very fact-based. Intelligent SJs can often 'hoover up' facts with alacrity. They may also have very good memories and may use their considerable recall of facts as a source of power to bully others.

It is entirely possible to hold high office and yet have a significant personality disorder. Peter Mandelson was a Cabinet minister and an adviser to Prime Minister Tony Blair until 2001. Just after he had been sacked from the Cabinet for the second time, the medical correspondent of *The Times* described him as having a textbook personality disorder. Such people[22] are all too often those who succeed and who will later crowd together, not always amicably, on the green benches of the House of Commons, will dominate the senior messes in the Armed Forces, and control boardrooms. [They] . . . are prepared to exploit others, to plot and plan and to work all hours of the day and night to advance their chosen causes so that favoured institutions, and they with them, will become powerful, controlling and successful. They crave respect rather than love, and although often admired, few are personally popular. In short, there is actually something wrong with them, but that doesn't stop them achieving high positions. By extension, extreme authoritarians might thrive in peacetime armies, and only be discovered in war.

Research into the personalities of peacetime senior British Army officers suggests a dichotomy amongst senior officers between authoritarians and autocrats. Further evidence indicates a possible basis for that dichotomy in terms of MBTI typology. Richard Sale, a former army officer who later taught at Lancaster University, conducted a psychometric study of senior British Army officers in 1989–90. His purpose was to establish whether psychometric tests could identify those officers who are selected for high command in peacetime. He concluded that, in simple terms, they can identify them.[23] His data also reveals other interesting findings.

Sale conducted eight tests, which provides scores on 36 separate personal characteristics. His 49 subjects were all serving colonels or brigadiers. They were commanding brigades or equivalent formations, or had recently been selected to do so. He divided his subjects into an elite group (of 20) of those chosen to command regular armoured, parachute or infantry brigades, or to attend the Higher Command and Staff Course. The remaining subjects were described as the 'pool'. Thus all had reached reasonably high rank, but the 'elite' group had been provisionally selected for very high rank.[24] Sale compared his findings with successful groups taken from civilian life: senior managers, senior brokers; and managers of surveyors, retail and sales personnel.

The 36 characteristics are not well described in Sale's work. They are only really meaningful to psychologists. However, the combinations of characteristics, and their comparisons with civilian groups, are meaningful to the non-specialist. The most important finding is that authoritarian tendencies are strongly prevalent in both the pool and elite groups. Authoritarianism is more marked in the army groups than any civilian comparator. It is even more marked in the 'pool' than the 'elite'.[25] Many of the senior officers surveyed show authoritarian tendencies. However, some of the elite show forceful, autocratic behaviour instead. Sale's statistics can be used to construct an 'Index of Authoritarianism' as displayed in Figure 24.[26]

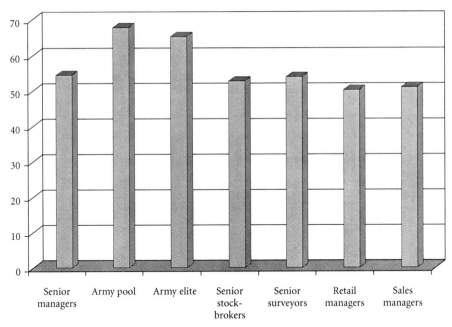

Figure 24: Index of Authoritarianism

Amongst the elite there was a dichotomy. Some of them were highly authoritarian, but the presence of some extremely autocratic individuals reduced the overall authoritarian character of that group. The pool were even more authoritarian.

The data show that senior army officers are considerably more likely to be authoritarian than their civilian peers. Sale also found evidence of neuroticism amongst his subjects.[27] That is unusual, because psychologists do not normally connect authoritarianism with neuroticism,[28] but it is consistent with Dixon's work. Neuroticism is characterized by a tendency to experience (and hence display) a number of emotions, of which anger (bad temper) typified the bullying commanders in BAOR in the 1970s and 80s.

Sale found senior army officers to be the least imaginative of all comparators. Their scores for 'original thinking' were generally low. This is perhaps surprising, since the army values personal initiative. Alternatively, it may suggest why it is so highly prized: it seems to be rare. There was, however, a very wide spread of army scores: the widest for any characteristic. Some of them must have been very dull indeed, as scores for verbal reasoning (akin to IQ) were consistently the lowest of any of the groups compared.[29] This is surprising, since the army has the most stringently applied entry qualifications.[30] It seems the army does not retain the brightest brains. Many of the army's better colonels were not as bright as a group of shop managers. Nevertheless, two of the elite group achieved outstanding scores for verbal reasoning. Other data suggests that they were highly autocratic. This reinforces the perception that the elite group are split between authoritarian and autocratic personalities. It also appears that authoritarian behaviour might mask low levels of intellectual ability. This seems to mirror previous remarks about the army not needing brilliant people. These findings are shown graphically in Figure 25.[31]

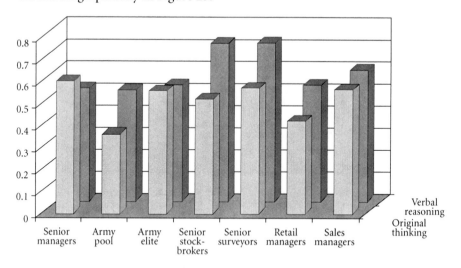

25: Original Thinking and Verbal Reasoning

Senior army officers scored the highest of any group for the characteristics of 'Personal Relations' and 'Sociability'. They appear to be 'people people'. Sale considered that a senior officer's 'only threat is that of being assessed by his superiors as being unfit for further advancement; it is therefore hardly surprising to find a strong emphasis on Personal Relations and Sociability'.[32] This could be a focus for authoritarians' deep-seated need to be loved, directed towards their superiors. It could also reflect the fact that the army has a strong 'officers' mess' culture.

Sale's aim was to create a profile of successful senior army officers. He constructed a profile of the 36 characteristics. He then scored all his 49 subjects against that profile. Most members of the elite group matched his profile, but six did not; some being seriously mismatched. There was no single characteristic common to all of those six, but they often gained an extreme mark in one area. It might be a very low score in characteristics which included original thinking and vigour. Alternatively, it might be extremely high in responsibility and cautiousness. These people would appear dull, dutiful or unimaginative; although possibly sociable, personable and forceful.

Overall, it seems that that senior army officers include the most authoritarian, cautious, controlling personalities of any group recorded. Luckily, the army's elite also includes some of the most daring, risk-taking and intelligent of all surveyed.

Sale's data refers to an important sample group at one point in time. In a similar study, over 85 per cent of all students of one year at the USMC Command and General Staff College were found to be SJs, despite Guardians being only 40 per cent of the population.[33] I have found no data for the British Army as a whole, but I did find an interesting analogy. Dental records of officer cadets at Sandhurst show very narrow variance.[34] This indicates remarkably consistent socioeconomic, cultural and physio-sociological background.[35] That does not indicate similar temperament. It is, however, a strong indication of effective pre-selection or self-selection amongst officer entrants. If it applies to health, it may apply to temperament.

Colonel Philip Barry conducted a study of typology amongst British Army Staff College students in 1999–2000.[36] Staff College graduates generally contain most of an army's future generals. The results are summarized by temperament in Table 5.

Table 5 Distribution of Temperament among British Staff College Students

Temperament	Title	Proportion of Staff College Students	Proportion of Whole Population
NF	Idealists	8.3 per cent	10 per cent
NT	Rationals	39.5 per cent	10 per cent
SJ	Guardians	43.4 per cent	40 per cent
SP	Artisans	10.5 per cent	40 per cent

The most striking aspect of these figures is the over-representation of NT Rationals. This contrasts starkly with USMC Staff College students, who were overwhelmingly SJ Guardians. Guardians are found in British Staff College students roughly in proportion to their occurrence in the population as a whole. Conversely, NTs are over-represented. In particular, ENTJ Field Marshals appear six to seven times more often than in the population as a whole. The commonest type overall was ESTJ Supervisor, with 22.3 per cent of Staff College students.

There is an interesting correlation with observed practice. The British Army training Unit at Suffield (BATUS) in Canada is one of the few places where the army regularly exercises at battlegroup and formation level. Over two years perhaps 15 or 20 commanders would have been exercised at those levels. In 2003 the outgoing commander of BATUS observed that only about 20 per cent of the commanders he had observed could form and hold a mental picture of a mobile armoured battle in their heads.[37] It is reasonable to assume that the 20 per cent would be NT Rationals, and that all were Staff College graduates. This may imply that Rationals are being de-selected between Staff College and unit or formation command, presumably in favour of SJ Guardians. Alternatively it could be that only half of the 40 per cent or so Rationals who go to Staff College can actually visualize a mobile battle.

Thinkers (T types) constitute roughly 50 per cent of the population, but 92 per cent of Barry's subjects. In simple terms, whilst USMC Staff College graduates were predominantly SJ Guardians, British Army Staff College students were overwhelmingly Thinkers. About 38 per cent of Barry's sample were introverts. He found that surprising. It undermines the stereotype of the charismatic extrovert senior officer. Extroverts can be disadvantaged in mental pursuits such as staff work, because without external stimulation they can find it difficult to remain focused on the task at hand.[38] That partly explains why many intellectuals are introverts. Thus the existence of introverts amongst those selected for staff training is useful. Had that not been the case, Montgomery might never have reached high rank.[39] This statistic also supports the notion that introverts are not necessarily shy. British Army officers may be many things, but they tend to be well socialized and not shy.

The main conclusion from Barry's data is simple: the emergence of leaders. ESTJ Supervisors and ENTJ Field Marshals were the commonest types present, and were strongly over-represented. Both types are described as 'natural leaders'. They are, however, quite different. ESTJs tend to be decisive, direct, efficient, responsible and task-focused. Their world is one of structure and plans. ENTJs are also decisive, but challenging, energetic, strategic and tough-minded as well. They will tend to be conceptual thinkers, while ESTJs will tend to focus on detail.[40,41]

To summarize: Sale's data show evidence of polarization in his sample of senior British Army officers. Some of his elite group were outstanding. They were highly intelligent, rational and lateral thinkers: amongst the best in any group surveyed. They can be seen as autocratic. The remainder showed a strong tendency towards authoritarianism. They were not particularly intelligent. Their behaviour was a mask for a lack of intelligence in some cases. Some of them tended to be hardworking, dutiful and responsible; possibly personable, but dull.

Barry's data, of a group 20 years younger and 10 years junior to Sale's sample, shows a pattern in which the Thinker type predominates. Over 80 per cent are either NT Rationals or SJ Guardians. This apparent dichotomy is modified by the fact that over 35 per cent are ESTJs or ENTJs, compared with less than 15 per cent in the population as a whole. They are separated only by their S/N score. Given that there is a sliding scale between extreme Ss and extreme Ns, with many individuals lying towards the middle, this dichotomy may not be very great. Nonetheless, it does appear that some extreme ESTJs are highly authoritarian. Conversely, some aspects of the behaviour of extreme ENTJs could be seen as autocratic.

Dixon's and Gordon's work both indicate two types of senior commanders: authoritarians and autocrats, regulators and ratcatchers. They suggest that authoritarians or regulators will tend to fail in war, whilst autocrats or ratcatchers will tend to thrive. Combat is dynamic, stressful, complex and confusing. The way a commander responds will be significant. S types, given their focus on the real world, will tend to concentrate on detail, whilst 'N's will tend to concentrate on patterns and concepts. Both ESTJs and ENTJs will tend to be decisive. They may both also appear charismatic.

ESTJs' main fault is that they tend to decide too quickly. They may not see the need for change (hence be conservative), overlook interpersonal niceties, and be overtaken by their emotions when they ignore their feelings for too long. Hence the anger and bullying. Conversely, although ENTJs might also tend to decide too quickly, their main fault is that they may overlook others' needs and contributions in focusing on a task. They may overlook pragmatic constraints, and they may ignore or suppress their own and others' feelings.[42]

People tend to revert to type under stress. Thus the worst aspects of a commander's temperament or type may present themselves in combat. SJs, and particularly ESTJs, may fuss over detail, micromanage, decide too soon and be unpleasant bullies. NTs, and particularly ENTJs, will be able to see the whole problem rather better. They may well be forceful, perhaps unpleasantly so, and hasty at times. Overall, the differences between regulating authoritarians and ratcatching autocrats may be reflected by the subtle but important differences between SJs and NTs.

This explains another apparent paradox: that successful army officers want to lead others and take charge, but also want to be led.[43] In practice two groups of people do resolve this paradox. One group is driven by status and conformity, which includes the need to 'kow-tow' to superiors. Authoritarians can be notoriously subservient (and the most effective way to deal with a bully is to bully him back). The second group, the autocrats, will be far more effective in war. They can see both how to make sense of the complex, dynamic, unstructured mess called battle and how to make their way through a peacetime hierarchy.

Thus we come to the crux of the problem. Stable hierarchies, such as armies in peacetime, will tend to attract SJs, and there will be a number of authoritarians amongst them. SJs are adept at learning the rules of the hierarchy, not least those of interpersonal behaviour, and they will be disposed to do so if they are extroverts. Unfortunately, authoritarians will tend to thrive in peace. The results of 40 or so years of such processes can be seen in Sale's results for 1989–90. It is hard to envisage a better example of a stable peacetime military environment promoting its own, and producing a large number of authoritarians.

Sale demonstrated that, since the army's annual confidential reporting system is entirely subjective, and promotion is based entirely on assessment by senior officers who are themselves the products of the system, promotion within the British Army in peacetime accurately reflects the army's culture.[44] Unfortunately, history suggests that many of those it promotes are found wanting in wartime, and that they are replaced by younger men who display other, more useful capabilities *in war*.

We have seen that armies can lose collective performance very quickly: perhaps within a year or two. Individual experience is lost more slowly, over a period of about 30 years. We have just discussed an apparent polarity between autocrats and authoritarians amongst senior British officers. We have also seen that ESTJ Supervisors and ENTJ Field Marshals tend to be over-represented amongst British Staff College students. Therefore we can very tentatively suggest that there is some correlation between some aspect of extreme SJ behaviour and authoritarianism; and also between aspects of extreme NT behaviour and autocratic behaviour. We can also suggest that the best available (the autocrats) can succeed to the higher ranks of the army in peacetime. Unfortunately so also do larger numbers of authoritarians, who are predisposed to fail the test of war.

These suggestions, or hypotheses, must be very tentative. Firstly, the data comes from four separate points over a century: senior commanders in the Boer War; their successors at the beginning of the Second World War; colonels and brigadiers in the late 1980s; and Staff College students in the late 1990s. The latter sample constituted only half of the available population and there may be a systemic skew in the data. For example, a certain type might be predisposed not to return the questionnaires.

Furthermore, there are acknowledged limits to the general state of personality theory. Not least, MBTI has limitations in its validity and repeatability as a test. A psychologist would not generally use it.[45] Barry concluded that it should not be used for the selection of senior officers.[46] MBTI is, however, indicative. Neither Dixon's nor Sale's work relied on it.

The British major-general Christopher Elliot once remarked that he had noticed a marked difference in the personality of the army's brigade commanders during the 1990s. At the end of the Cold War they tended to be elegant, sophisticated, personable, and often cavalrymen (or gunners). By 2000 they tended to be gritty, often northern, infantrymen and sappers.[47] He attributed the difference solely to the army's changing role.

General Elliot suggested that BAOR required presentational and social skills, supporting and maintaining the appearance of deterrence over decades. That environment gave way during the 1990s to frequent operational deployments to a number of theatres including the Gulf, Bosnia, East Timor, Sierra Leone and Kosovo – all within ten years. This shift from presentation to practicality may reflect pragmatism in the selection of senior officers. We should be sceptical: if an army promotes a certain type for long enough, eventually those doing the selecting tend to be of that type. A simpler explanation may be more appropriate: those who succeeded during a long period of stability were found wanting once the situation became more complex and dynamic. It would be interesting to repeat Sale's tests on a current cohort of brigade commanders. It would also be instructive to compare the results with a sample of Sandhurst cadets; and perhaps also to track a cohort of Sandhurst cadets over 35 years. Such longitudinal surveys are expensive, but sometimes indispensable.[48]

There are possibly two other related factors. Firstly, one would expect authoritarians to be status-conscious, and reflect this in their choice of regiment or corps on joining the army. They would tend to join regiments which they perceived as 'smart'. This would tend to suggest a prevalence of authoritarians amongst the Household Division, the Cavalry and the Royal Horse Artillery (RHA). By coincidence, armoured (and hence Cavalry) and artillery (hence RHA) units spent long tours in BAOR. This may reflect General Elliot's comments about elegant cavalrymen and gunners. There also appears to be very little real correlation between 'smartness' and operational efficiency. If anything there is anecdotal evidence to the contrary, and other indications that smart regiments are simply good at self-promotion.

The second factor is education. The 1970s marked a gradual transition from non-graduate towards graduate officer entrants.[49] The '30-Year Rule' suggests that colonels and brigadiers would reflect that change 20 years later. General Elliot's elegant cavalrymen of the early 1990s are roughly Sale's cohort of authoritarians, recruited in the late 1960s and early 1970s. The 'gritty northern infantrymen and sappers' might reflect a generation of northern grammar school boys who went to university and subsequently became brigadiers. When graduates were in a relative minority, it is possible that graduates who joined the army were less likely to be authoritarian than non-graduate entrants, given authoritarians' disinclination towards intellectual activity.

Such observations should be considered with some scepticism. Insufficient data means that these observations can be little more than conjecture. However, they support the overall thrust of this book. Kuhn suggested that new theories will tend to gain acceptance if they describe useful phenomena, predict where other relevant phenomena may be found and possibly predict why such phenomena occur. Combat, and hence war, is fundamentally human. We have now made some observations about armies as human institutions, and suggested why some of them occur. We have observed that the emergence of authoritarians in peacetime armies is a brake on operational effectiveness in war. We have investigated what constitutes an authoritarian personality and how authoritarians succeed in stable hierarchies. In this instance, the ideas developed within this book appear to have pragmatic value.

10

The Human Face of War

In this book I have painted a depressing picture of armies forgetting old lessons and of authoritarian senior commanders. To that should be added a tendency not to improve in peacetime, or be distracted by the tactical lessons of the most recent counter-insurgency campaign. Not surprisingly, a persistent criticism has been of commanders preparing to fight the first battle of the next war the same way as the last battle of the previous one.[1] How can we avoid this? How should a nation build a better army over the long term? The final chapter addresses these questions.

There is historically little experience of a major army developing significantly in peacetime without a major and normally detrimental experience to spur it on. In 1806 the Prussian army was living off the reputation it had earned under Frederick the Great 40 years earlier. At the battles of Jena and Auerstadt Napoleon's armies inflicted shattering defeats on Frederick's successors, leading to the humiliation of the Prussian army and state. The senior figures who came through that salutary experience – including Scharnhorst, Gneisenau and Clausewitz – went to great lengths to reform the Prussian army. Those whom they taught, including Moltke the Elder, carried on that process. Two significant human issues cemented it in place. The first was Clausewitz's perception and genius. The second was that Moltke the Elder was Chief of the Prussian (and then German) General Staff for 30 years. The conditions which allowed Moltke to develop the German armies continually over several decades is unparalleled in modern history.

Thus it appears that without a recent and typically detrimental experience to spur them on, major armies rarely make significant qualitative improvements in peacetime. Why is that? Three reasons can be suggested. One is the human issue of institutional conservatism. This is simplistic: armies can and do change rapidly in war, once presented with evidence of their shortcomings. Such change may, however, be no more than a temporary sticking-plaster, and the immediate lessons of that war may themselves ossify during the next peace. The second possible reason is cost. Changing an army costs money. However, this is also simplistic: if most of the important issues are human, the monetary costs of changing human institutions can be relatively small. The third possible reason is the analogy of Gothic cathedrals. In the absence of a valid paradigm, the development of cathedrals in Western Europe took centuries and was marked with numerous failures: naves, facades or towers that collapsed. Thanks, however, to the discovery of Newtonian physics and an understanding of the science of new materials, by the nineteenth and twentieth centuries man could build structures which were many times longer or higher, and sometimes both, than those cathedrals. They also became safer. Or look at the development of passenger aircraft between 1945 and 1970 (by which time both Concorde and the Boeing 747 were in service). In both cases a common, agreed understanding of the underlying principles allowed very rapid development. Given a common, agreed understanding of the underlying nature of conflict and the behavioural human sciences, armies should be able to develop rapidly and continually for long periods.

One might think that in considering discussing the army of the future we should dwell on its size and shape. We won't take that path, because the size and shape of an army is generally dictated by issues of governmental policy, strategic situation, and so on. However, a few remarks are appropriate. We looked in detail at the structure of armoured divisions. We discussed issues such as flexibility and agility which emerged from seeing combat as complex, dynamic, adversarial, human, and so on. Yet, in most Western armies, armoured divisions are not like the 'lean and mean' formations of the Second World War. Current US, German and British divisions are huge. Why?

The real reasons for the size and shape of current formations have little to do with combat effectiveness. In the case of the British Army the overall shape and size is based on the 'Option Whiskey' (that is, 'W' as opposed to 'X' or 'Y') establishment that was chosen at the end of the Cold War. Much of what drove Option Whiskey was the requirement to make it as robust as possible in order to keep the army as large as possible. For example, the Royal Engineers managed to create a case for an engineer battalion in each armoured or mechanized brigade, thus guaranteeing at least six engineer battalions in the army overall.

There is an oft-repeated argument that unless something is permanently established within a divisional establishment it will disappear in the next round of cuts. The net result is simply divisions that are not fit for purpose: they are too big and too unwieldy, and are never used as intended. Let us be more honest. We might need large numbers of logisticians, engineers, medics or whoever because of the enduring requirements of peacekeeping operations. Alternatively, we might need them for a full-scale war. If we don't, or can't make that case, let us use the money for other things. In practice the current size and shape of Western armies reflects issues that are not primarily related to warfighting effectiveness.

Doctrine is a relatively new construct for English-speaking armies. Many armies have had training pamphlets and manuals for decades, if not centuries. However, for English-speaking armies the idea of an overarching description of what an army is for, how it operates and why it is organized the way it is, is fairly recent. Britain started on that path in 1989 with the publication of *The British Military Doctrine*. It is not true that Britain had no doctrine prior to that. It had no *coherent, explicit* universal doctrine. It 'just knew what to do'. Field Marshal Bagnall was both the first British officer to command an armoured division who was not commissioned during the Second World War, and the individual responsible for the writing of *The British Military Doctrine*. The generation of officers who joined with and after him did not 'know what to do' from first-hand wartime experience.

Thus we should discriminate between implicit and explicit doctrine. Many armies have implicit doctrine. Their officers share a common view of what to do, which is not written down and made explicit. It is based on 'the way things are done' and which are taught at officer academies and Staff Colleges, which is in turn based on 'the way things have been done'. Unfortunately, there is no way of knowing whether implicit doctrine is genuinely common across an army. Terminology will tend to be imprecise in all but trivial things: until the 1990s everyone in the British Army knew what a quartermaster was, but could not explain operations based on manoeuvre as opposed to attrition. Furthermore, if doctrine is not explicit it may drift over a period of years, and no one could say whether it had drifted or not.

Doctrine is also, by definition, what is taught. Therefore it should be 'what is *to be* taught': it should be authoritative. To be genuinely common across an army it should be explicit. It must be several other things as well. It must be relevant: it should describe situations, and where appropriate prescribe actions, that are likely to be faced by the army within the foreseeable future. It should also be coherent: its parts must fit together as a uniform whole without inconsistencies. Incoherence is partly a product of an inadequate underlying paradigm. However, some incoherence can result from doctrine writers not doing their job well. I have seen some appallingly bad doctrine.

The body of doctrine should also be practical, so that it can be applied to practical battlefield situations within the capabilities of the forces to hand. It should be teachable: it should be written at the intellectual level of the students for whom it is written, and be understood by their teachers (which has, perhaps surprisingly, sometimes been a problem). 'Relevant' should also mean 'up-to-date'. If, say, some aspect of the doctrine of the Cold War is still relevant in the early twenty-first century, then its practitioners should understand why and how, or they will tend to see it as irrelevant and discard it. Thus doctrine should contain a body of thought that is authoritative, explicit, coherent, relevant, practical and teachable.

The apparently mundane details of personnel policy also merit attention. Team integrity is a huge and unexplored driver of collective performance. Major exercises typically happen perhaps once a year. If the members of a staff team are posted on every two years, then on every occasion half have never worked together before, and may never have actually performed their particular task before. The other half would have done so once. If they were posted every three years, the proportion of novices would be one-third, which is not much less. A similar number – a third – would have been through the cycle twice. Those with some experience would outnumber the novices by two to one. The social cohesion of the team would also be enhanced.

From 1998 to 2000 my battalion served in a specialized FIBUA role in Berlin. By chance the key officers in battalion headquarters stayed in post for virtually the whole tour. That was very unusual in the British Army. Within a few months of arrival in Berlin, they had mastered the intricacies of controlling a battalion in FIBUA, and the battalion's collective performance continued to improve throughout the tour. From Berlin I was posted to HQ British Forces, Falkland Islands. We ran an exercise for all troops in theatre every month. Nonetheless, collective performance never improved – largely because of high turnover due to very short tours. On every occasion similar problems were identified, but in different parts of the command. These examples illustrate the systemic, human effects of continuity on collective performance.

Why are tours of duty not longer? The fundamental answer appears to be human and structural. The British Army's Military Secretary's branch manages careers and career structures. In practice it is focused on the people. Much 'personnel management' appears to be focused on 'career development'. Units and formations are the responsibility of their respective commanders. They are posted in, serve, and move on. There is no branch of the army which is systemically responsible for the best interests of the units or formations. 'The interests of the Service' are largely expressed in terms of filling posts. There is no corporate voice which says that the unit or formation would be better served with longer postings. Whenever such suggestions have been made (normally citing advantages in terms of reducing personal and family disruption), the prevailing arguments are those of the difficulties of career development. Systemically, the interests of the individual prevail over operational effectiveness, because the latter has no institutional voice.

Fostering innovation is a further aspect. The American analysts Anthony Cordesman and Abraham Wagner consider that innovation, training and manoeuvre are all part of one process. That process should be encouraged as much as possible, and tied to operational needs in order to maximize effectiveness.[2] Innovation is essential to an empirical, pragmatic approach. However, it seems that high officer rotation promotes innovation, but only in relatively flexible organizations. Thus the combination of innovation (which requires high officer rotation) coupled with high collective performance (which requires continuity) seems highly beneficial, but unachievable. However, what constitutes 'high officer rotation'? Tours of duty of four to five years in a unit are apparently the norm in the Israeli army.[3] It is not clear whether 'high officer rotation' means 'in one unit' or 'in one post'. If the latter, the Israelis would have astonishingly *low* rotation by British, and particularly US, standards.

Personnel policies and issues have a major effect on command and collective performance. There are numerous examples from the Vietnam War; mostly of bad personnel policies crippling operational effectiveness. The '365-day' syndrome and high rotation of officers both resulted in very high rotation of personnel within units: most officers served only three months in post. The 'shake and bake sergeant' syndrome resulted from the requirement to produce numbers of NCOs from conscripts: there were insufficient regular NCOs for the task. Other examples include the attempt to form a second contingent of airborne forces, once the first contingent had served one tour of duty; the politically driven enlargement of the army without a sufficient cadre of officers and NCOs; and the woefully low standards of training, discipline and morale that resulted.[4] One feels little less than pity for much of the US Army in Vietnam. The similarly organized and equipped, but substantially regular, Australian army initiated contact with the enemy on average seven times as often (in 84 per cent versus 12 per cent of all engagements), and suffered significantly fewer dead as a proportion of the force.[5]

Experience is the best way to achieve practical coordination and overcome the fog of war,[6, 7] as long as the experience gained is positive. Collective human experience is highly dependent on team integrity. If the team is kept together for long periods it can learn: both directly, and in its ability to learn. Patton was adamant that divisions should be fought as entities wherever possible.[8] Conversely, the British tactic of massing armour in Normandy in order to avoid infantry casualties is reckoned to have had an adverse effect on the performance of the divisions from which the armour was drawn.[9]

During the First World War British staff came to realize that 400 lbs of high explosive was required to neutralize each yard of trench.[10] The US Marine Corps' landing at Inchon in Korea had been thought impractical, but MacArthur's planners included a large number of highly experienced Marines from the Second World War. They knew both the problems of amphibious warfare and the practical ways of overcoming them.[11] Both of these examples relate to highly pragmatic facts learnt from experience and observation. Facts can be recorded. Intangible aspects are more problematic. They are more likely to be forgotten in long periods of peace. Here we will consider how forces learn, rather than how they forget.

Armed forces do learn. Indeed, the performance of many formations in combat is largely understandable once their relative experience is accounted for. The example of the US 29th Infantry Division on Omaha Beach was discussed earlier. It is noteworthy that the experienced 1st US Infantry Division also assaulted Omaha Beach on D-Day but did not suffer as badly as the green 29th Division.[12] Similarly, the success of the two first-echelon divisions on Operation Totalize in Normandy in 1944 – the (by then) veteran 2nd Canadian and 51st Highland Divisions – can be contrasted with the relative failure of the 'virgin' 1st Polish and 4th Canadian Armoured Divisions of the second echelon.[13] Experience is clearly a factor in explaining such differences.

The Wehrmacht refused to reduce the length of officer training during the Second World War, despite front-line officer shortages.[14] It was short of 12,887 officers in December 1944 (at which point the US Army had 50,000 officers surplus to establishment).[15] It appears that the Wehrmacht knew that poorly trained officers simply led to higher casualties, and that it was better to make do with fewer than put up with poorer officers. This is interesting. First, it was common wisdom in the Wehrmacht. It must have been identified in, or in the aftermath of, the First World War. Secondly, as Generals Beck, Halder, Zeitzler and Guderian all defied Hitler on the issue, they obviously felt strongly about it.[16] This suggests an understanding of the connection between command effectiveness and collective performance which was missing in Western armies.

An obvious inference is that officer training and education are more of a factor in operational effectiveness than is commonly suspected, not least, in units' and formations' ability to learn. However, the impact of changes to officer training policy are difficult to identify until years later, and so the linkage may not be obvious. There is also a tendency to equate ability with rank and therefore increase the rank held by a certain post over time. During the Second World War British divisional commanders' engineer advisers were lieutenant-colonels, and brigade commanders were advised by majors. By 2000, divisional commanders were advised by colonels, and a brigade commander's engineer adviser would be a lieutenant-colonel. Why? In the German army he would be a captain. Surely the issue is that he is selected and trained to do the job, and has the authority to do so. There does seem to be a need for a more senior officer, for example, in a brigade-sized expeditionary deployment. But that is the case for the rank of the expeditionary force engineer adviser, not for every brigade engineer commander.

Individuals should be given the opportunity to learn from experience in junior ranks. If a job that can be done by a captain or major is done as a matter of course by a lieutenant-colonel, then the captains and majors will not get much experience working for him. When they become lieutenant-colonels, they will have less experience than they would otherwise have had as captains and majors.

They will also probably become accustomed to referring decisions upwards, and hence display little initiative. Such policies appear locally sensible but are in the long term self-defeating. The alternative entails some risk. An unsupervised junior officer may make mistakes based on inexperience. However, the issue is like that of mission command, of learning to lean downhill when skiing. It may take some confidence at first, but should become natural in due course.

It is critically important to train staffs.[17] While most armies have some form of individual staff training, it seems that in practice few staffs are given the opportunity to train themselves, rather than be exercised. When Rommel was appointed to command Panzer Gruppe Afrika, he was allocated a staff which was given a month to train together in Bavaria. It certainly contained some gifted individuals, such as Siegfried von Westphal and Friedrich von Mellenthin, both of whom went on to achieve high rank. However, it seems that much of its effectiveness came from its collective training. Only four of the staff were Staff College graduates.

There seems to be a huge gearing effect. With better staff, and well-trained and educated officers, a formation is optimally placed to fight and win its first battles, and then learn from its experience. Not least, it can be expected to have an effective process to capture lessons and incorporate them into practice. Furthermore, a trained formation with a trained staff can produce shorter orders and act effectively on them. Those orders can be produced more rapidly. Hence the formation can achieve greater tempo, making it more effective. The gearing is on several levels.

It is clearly important for an army to have a suitable process to select the next generation of commanders and senior staff officers. There is a tension here. Since no army gets paid to come second, selectors should pick those who will do best if, and when, the next conflict occurs. Yet each nation will have cultural expectations as to who should be promoted. In Britain one of the parameters seems to be the semblance of fairness: everyone should have an equal chance based on merit and potential. In some armies there is an expectation that those of a certain caste, or those related to the national president, will be preferred in practice. Many of Saddam Hussein's senior commanders came from his home town of Tikrit. Would Tikritis be the best in war? More subtly, will those chosen primarily because the system is seen to be fair inevitably be the best in war? In practice the dynamics are not obvious. Two examples, taken from real life in the British Army, demonstrate the point.

Some years ago an infantry battalion had three rifle company commanders. By chance they all served together for almost all of their two-year tours of duty. The first was a Staff College graduate. He was very focused on training his company for operations. His company was very good. When the battalion underwent training for peace support operations, the external training team said that one of its operations was the best it had ever seen. On the last occasion that the battalion trained together, the divisional commander was told by the CO that it was 'his best company'. The second company commander wasn't a Staff College graduate but was a very good trainer. He was popular with his soldiers, and they would do anything he asked of them. The third company commander was a Staff College graduate, dedicated and professional, but somehow his company never really shone in comparison with the other two. The three OCs avoided any appearance of competition between them, yet generally the first company was generally better than the second, and both outshone the third. So we have three company commanders given a fairly equal chance of training and preparing their companies for war. One became the CO in due course. Who? The third.

In another brigade and on another occasion there were three COs. One was brilliant. This was during the Cold War, and the brigade and its units trained and exercised frequently in north Germany. His unit really was outstanding. The corps commander, subsequently the Chief of the General Staff, said in public that the battalion was the best battlegroup in the British Army of the Rhine (out of a total of 26). That was a considerable accolade: the unit had been in Germany for just 24 months. The CO rose to the rank of colonel, but was only promoted in time to get a colonel's pension on retirement. He was the only senior officer in the brigade not to reach at least the rank of major-general. British Army officers reading this will say that of course there are many factors other than tactical ability that influence promotion.

Quite so. As the two examples show, on whatever basis the British Army actually promotes to high command in peacetime, there can be absolutely no doubt that tactical ability displayed in training and on exercises is not a major factor. In practice, like other armies, the British Army has a set of implicit standards which are the real, effective criteria for selection and promotion. At captain and major level they have been called the 'pretty adjutant syndrome'.[18] Does the army select and promote on the basis of being tall, good-looking, well-spoken and self-confident? There are other, more objective, examples.

The 'Junior Officers' Training and Education System' promotion and staff-selection test, introduced in 1988, is one. Three years after its introduction a letter from Infantry HQ suggested that 'the wrong people are passing'. In practice it meant that 'the right people aren't passing'. That was despite the fact that the new arrangements were demonstrably more objective than their predecessors. The staff selection test was replaced by a system which gave more weight to subjective annual confidential reports. In effect objectivity was beaten by military culture.

We have taken a short look at selection, based on a few examples. Clearly the reality is far more complex. But let us be quite clear about the importance of the issue. As we have seen, only about 40 per cent of a typical tactical decision comes from the information presented. The other 60 per cent is brought to the decision by the decision-maker, primarily in terms of his personality and experience.

Have armies invested as much in selection, training and education as they have in the electronic communications and information systems which deliver the information?

The need to avoid extreme authoritarians reaching high rank is another issue. That is not because they're not nice people, but because they will fail in war. A nation does not pay its army to come second. There is a real problem here. Armies need rules for their selection procedures. Unfortunately authoritarians will be adept at learning, taking advantage of and even manipulating those rules. In fact those rules more than any other, since selection relates to advancement, and advancement to status. So, in an apparent paradox, we want a set of rules which disadvantage those best suited to taking advantage of them! The generational effect or '30-year rule' is another problem. Changes to selection rules may take 30 years either to produce the required results, or to reveal problems. With a 30-year feedback loop, the selection of senior officers will not be a precise science.

Would objective psychometric testing be of value? I have spoken to several psychologists with first-hand knowledge of the British Army. They all said that they would be most surprised if it accepted psychometric testing as a major part of senior officer selection. The reason is not easy to ascertain. The army appears to reject explicit testing, at least amongst its middle- and senior-ranking officers. Such tests are not seen to be fair. Note that the perception of fairness is being preferred to objectivity. This appears to be a deeply held belief, hence deep in the army's culture and hard to change. That is not to say that it can't, won't or shouldn't change. But resistance should be expected. We have dwelt on the British Army here; similar cultural considerations will apply in every army.

We should not be too pessimistic. Western armies are generally quite well run, and successful on operations. For example, the British Army does seem to contain several of the rational, empirical brains needed in war, and some superb autocratic leaders. Yet it does not necessarily select the best in peacetime. Changing that would require some relatively direct and concrete measures. It would also require deeper, institutional and organizational changes which would impinge on its values and beliefs. Such cultural change would not be made easily.

We remarked earlier that armies rarely seem to learn in peacetime without some recent and generally negative operational experience to spur them on. Ideally there should be a constant search for improvement. This applies as much to the individuals as to the institution as a whole, and it does not only refer to officers. During the Edwardian period the British Army invested heavily in education for its rank-and-file soldiers, and gave financial incentives for gaining educational qualifications. Perhaps surprisingly, the number of courts martial went down significantly.

If the army of the future is to employ its empirical and pragmatic brains more effectively, its key practitioners should gain the habit of learning throughout their careers. Today Continuing Professional Development (CPD) is a fundamental element of many professions. A barrister is obliged to keep abreast of the latest in case law, a doctor to keep up with *The Lancet* or the *British Medical Journal*, and so on. In many of the self-regulating professions today, CPD is more than that. It generally requires annual evidence that the individual is maintaining and developing his fitness to practise.

However, British Army officers seem to resist any requirement to undertake private study. They seem to consider that long residential courses should teach them all they require for a job. They are perhaps right in the narrow sense of 'teaching'. They also resist out-of-hours study, citing (perhaps reasonably) that the requirement to do so often falls when some of the relevant cohort are deployed on operations, putting them at a systemic disadvantage in comparison with their peers. That may be so. Part of the issue is that private study seems only to be linked with competitive examination. There may, however, be another issue. Authoritarians are disinclined to reflect, and Sale's subjects showed a deficit of reasoning skills. CPD might have the effect of deterring authoritarians from persisting with a military career.

The general effect of introducing CPD should be that, at each point in history, the current cadre of officers (encouraged by their superiors) should aspire to be better than their predecessors. 'Better' implies better-educated throughout their career. They should also consciously improve and develop the organization for their successors. A key tenet of the Prussian General Staff of the early nineteenth century was that its members were self-consciously the guardians and developers of the system that had nurtured them.[19] During 25 years' commissioned service I was never presented with a view of the future of the British officer corps which was different from the status quo, let alone better.

In war, winning is all-important, and each army should strive to be the best. Surely producing the best is more important than a perception of fairness in the selection process? This logic extends to generating an elite. Members of elite are, by definition, the chosen ones; by assumption, the best. This is not the same as encouraging everyone to do the best they can (which is muddy thinking). If possible, both should be sought. The army of the future should consciously select, train and educate those best capable of delivering success on operations.

Whenever a war or major operation occurs, the best available commanders and principal staff officers should be in command. To prepare them for those appointments in a reasonable timeframe there is a fairly limited number of jobs they can do on the way up. So few, in fact, that they should concentrate almost exclusively on those jobs. For example, in the British Army from commanding a company in one's early thirties to reaching the rank of lieutenant-general at or about the age of 55 one should have commanded a battalion, brigade and division. That alone is over six of the roughly 20 years available. It is also desirable to have worked in brigade, division, corps and possibly higher-level headquarters. In practice the man best suited to commanding a corps at the age of 55 is likely to have held relatively few other than command and major operational staff appointments. He will not be a broadly based generalist in any recognized way, except that he can work across a number of staff disciplines within an HQ. And that is no bad thing: we want him to be the best, most experienced man for the job.

If the elite it is to be intellectually curious it should be well educated. Investment in education would be a significant step. It might, for example, be reasonable to give the best 10 per cent of officers an additional year's education: perhaps at university, at an overseas Staff College in addition to his own (as the Bundeswehr currently does), or on a major research project. The US armed forces place selected officers in 'think-tanks' for a year at a time. That would not be cheap, but would be an investment in excellence.

We started this book by looking at current military thought. We found that not many people do it, and what thought there is is often incoherent, inconsistent, and riddled with paradox. Much of it simply isn't very good. The OODA Loop is simplistic, whilst the Principle of Four falls apart as soon as one applies any rigour to it. The whole 'effects-based' debate of the first few years of the twenty-first century lacks any apparent consensus or direction of progress. And what happened to the idea of a Revolution in Military Affairs, much heralded in the previous decade?

In practice much military thought is not simply mythological. It often has some, possibly flawed, basis in reason and the needs of some pragmatic policy issue. In peacetime much of it is driven by internal political imperatives. The Principle of Four was useful in protecting the size and shape of the British Army in the later 1990s. The 'effects-based' discussion was useful at one point in the procurement of stealth-bombers. Such new ideas are often opposed by innate conservatism or downright pragmatism – even in a good army, navy or air force. Yet all these issues are human – psychological, social or cultural.

Much of the difficulty which 'effects-based' campaigners currently face is that they are trying to materially change thinking in an intellectual discipline where there is no mature paradigm. They don't seem to know that. They are trying to address a wide and diverse audience which has no shared view of the problem. Therefore there is no shared language and no mutual understanding of cause and effect. I suspect that they will ultimately fail, because there is no great tension or force driving acceptance of their views. An article in May 2008 suggested that the effects-based approach was being gradually accepted on the entirely pragmatic grounds that it provides useful linkage between military action and political guidance at the operational level.[20] That may be so. However, the article also highlighted intellectual inconsistencies between the effects-based approach and the technique of 'centre of gravity analysis'. This is a clear example of what we described earlier as 'dogma, or slabs of belief held together by their own self-contradiction'.

More generally, however, military thought is not well developed. We found much paradox, which prompted us to search for a new paradigm with which to resolve it. We seem to have gone a long way down that route. 'The attack is the best form of defence'? Largely because by attacking, one forces the defender to protect himself. That reduces his ability to inflict harm on his attacker. Other aspects of conventional military wisdom have also dropped into place. For example, high tempo is generally preferable to low because it prevents the enemy regaining the initiative and creates opportunities to inflict and exploit surprise.

After a detour through philosophy we developed an explicit paradigm for combat. We should base our understanding on a fundamental premise: that fighting battles is basically an assault on the enemy army as a human institution. Combat is an interaction between human organizations. It is adversarial, highly dynamic, complex and lethal. It is grounded in individual and collective human behaviour, and fought between organizations that are themselves complex. It is not determined, hence uncertain and evolutionary. Critically, and to an extent which we generally overlook, combat is fundamentally a human activity. It has a human face.

This should allow us to develop better theories, if the word 'theory' is not too grand. Take the notion of the Manoeuvrist Approach. After about ten years discussing it, the British Army came to the view that it was an approach based on attacking the enemy's will and cohesion; that seeks to win at least cost, pitting strength against weaknesses; and prefers original and indirect approaches to military situations. This is not cast in stone, nor dogma. It is a construct which, empirically and pragmatically, seemed to work. Moreover, statements such as 'armies don't get paid to come second', or that 'conflict is, fundamentally, resolved when the enemy believes himself beaten' may not seem like golden nuggets of essential truth. Nor are they intended to be sound-bites. However, they should now fit into an overall picture of what combat is like and how best to engage in it.

Our paradigm generally calls for tactics which attack the enemy's will and cohesion; which prefer infiltration and seeking flanks and rear; and systemically exploit the enemy's weaknesses in pursuit of the higher commander's intent. Surprise (doing something the enemy in practice does not expect) and shock (the sudden, concentrated application of violence) should be both inflicted and exploited. Our paradigm calls for organizations that are agile and flexible; that can react swiftly; that can gain and if necessary regain the initiative, and hence turn the tide of battle. Those forces should generally be organized into smaller packages than the leviathans of the late twentieth century. The paradigm calls for command posts that are small, mobile and contain small numbers of well-selected, relatively junior but highly trained staff. They should be generalists not specialists, and the emphasis should be on quality not rank. The paradigm calls for commanders who are capable of forming, maintaining and enunciating a vision of the big picture. They should be incisive, decisive and reasonably forceful people. In particular they should not be authoritarian bullies, nor those who thrive on the detail; least of all, both. And it calls for decisions that are 'about right', but which can be turned very quickly into concise and clear orders. It calls for a highly educated, self-improving officer corps that is institutionally aware of its responsibilities for the army of the future.

Few of the details contained in this book are critical in themselves. Fundamentally, what is needed is a philosophy of limited empiricism and pragmatism. Armies need to get away from reliance on methodologies of cause and effect; from technological paradigms; and from attempts to decompose what is in practice an indescribably complex problem. It's not so much that they won't work, but approaches which are holistic, systemic and heuristic are likely to work better. Even the notions of shock and surprise are, at most, probable hypotheses as to what is generally best.

As the last decade of the twentieth century closed and the next opened, there was a minor industry in articles about a revolution in military affairs. It appealed to two cohorts and was therefore a fruitful area of debate. On the one hand it allowed historians to debate the concept at length. The original idea, that Western powers had undergone such a revolution with the advent of gunpowder weapons, was reasonable in itself. But then we saw some historians claiming that there had been several such revolutions; or that the gunpowder revolution went on over a period of anything up to 400 years; and great debate as to when it actually occurred. Many historians had a field-day.

The second cohort were the technophiles. Driven by the fierce white heat of technology, and particularly the application of digital electronics on and over the battlefield, the world is apparently now experiencing a second (or third, or fourth, or 'another') such revolution. Imagine the appeal: a technologically forward-leaning idea with impeccable historical credentials. Well, perhaps: one great joy of history is that it is rarely categoric. And it attracted some fairly predictable support. If the procurement of sophisticated electronic weaponry had transformed the nature of war, then large and increasingly expensive amounts of sophisticated electronic weapons and sensors should be purchased in order to keep up. How wonderfully self-fulfilling.

Yet the reality of the next big war, in Iraq in 2003, wasn't like that. The army's analysis of the major combat operations that year concluded that

> The Coalition went rapidly to where the enemy was, surprised him, and then outfought him. There were parallels with Marlborough's campaign to Blenheim; Napoleon's Ulm Campaign; the fall of France in 1940, or the Sinai Campaign in the Six-Day War . . .
>
> British and US forces deployed swiftly and efficiently, and achieved the stated operational objectives equally swiftly and effectively. Notwithstanding the efforts and at times sheer courage of our armed forces, the defeat of an Iraqi regime that had been crippled militarily in 1991 and had no chance to reconstruct its forces in over a decade was not really in doubt.[21]

Many of the things which came to our television screens brought out the human face of war: not least the mud and misery inflicted by the weather late in the first week of the campaign. In addition:

> The terms 'Effects-Based Operations' (EBO) and 'Networked-Enabled Capability' (NEC) were not found in extant doctrine in March and April 2003. They were at most statements of policy, concepts or aspirations. Thus the use of the term 'effects-based' in connection with Operation TELIC is misleading . . . It is also unfortunate to see such terms paraded with the flimsiest of justification. For example, smart munitions are of themselves not network-enabled. To see statistics concerning the increased use of [precision-guided munitions] as evidence for the efficacy of NEC during Operation TELIC is inappropriate. Public reports of Operation TELIC have at times indulged in the over-enthusiastic use of such terms without proper justification. The risk is that such usage is subsequently used to support policy or doctrine, without a proper basis in observed fact.[22]

It is perhaps interesting that far less talk of a revolution in military affairs is heard today. One criticism is that RMA theory generally lacks a human dimension.[23] Armed forces are being 'transformed', better to exploit the advantages of new technology. That is reasonable. However, there seems to be less thought that war is determined by technology than there was a few years ago.

Indeed, we should be very doubtful that there will be any revolution in military affairs without a related revolution in military thought. We have suggested that combat, and hence war, is not determined to any useful extent. It certainly doesn't appear to be determined by technology alone. Indeed, we have seen in several places that adherence to a technological paradigm has lead to false or misleading results. A technophile might be disappointed to find little reference to technology in this book. Unfortunately technophiles tend to equate war with the use of equipment, and look to technology to solve all military challenges.[24]

Might we not expect some reference to technology in our paradigm? Well, try the converse. Is combat necessarily dominated by technology? Most unfortunately, it is not. There are too many conflicts around the world where the deployed levels of technology are low, unequal, or both. Clearly technology can create an advantage. However, in terms of our paradigm, we can see that to be a simple consequence of the adversarial and evolutionary nature of war. Armies will normally seek to develop and equip themselves with better weapons to give themselves an advantage over their opponents. However, we should always remember that 'the person behind the gun matters more than the gun itself'.[25]

Why, then, does technology figure so highly in common perception? First, it is visible and dramatic. A modern main battle tank firing accurately whilst moving at high speed, or a fighter ground-attack aircraft doing so even faster, makes better TV than a section of infantry trying to winkle out a sniper in downtown Baghdad, Basra or Kandahar. Secondly, it attracts money in large amounts. Thirdly, it does without doubt confer advantages: well-equipped high-technology armies have many advantages in circumstances where their technology can be deployed. But pause and look for a moment at the Vietnam War, which suggests that superior technology is not always the deciding factor.

Perhaps most importantly, people *believe* it gives a categoric advantage. 'Whatever happens, we have got – the Maxim Gun and they have not.' However, even this is probably a product of the current dominance of Western technology. There are relatively few examples of an enduring technological advantage, not least because war is adversarial and evolutionary. None of this suggests that technology is not a good thing. Of course it is. But its biggest weakness is that it implies a technological paradigm, which is flawed. War is not strongly determined, let alone determined by technology. Much of what is wrong with current military practice comes down to an assumption that more, and more detailed, planning is always a good thing. That would only be true if war were strongly determined. Unfortunately, it doesn't seem to be so.

What have we achieved? We have a book about military theory. It addresses some aspects of land warfare. It attempts to develop a new paradigm of land combat and hence some useful ideas about personnel, organization and tactics. It is neither 'right' nor 'wrong'. It doesn't attempt to be. It attempts to be useful. It will be valuable if it gives something of use to the practitioner. There is some, limited, evidence at time of writing that some of the ideas contained within it are helpful.[26]

However, military thought is not primarily about writing. Theory needs to be taught and taken into practice. In a sense it's like looking at a bolt of cloth; it might look smart or pretty, but it is not much use until it is made into clothes and worn. Military theory or doctrine can take years to have an effect. Some of the aspects of the *British Military Doctrine* of 1989 relating to the Manoeuvrist Approach were not fully enunciated until 1994, and were not fully established in practice by 2005. Some aspects probably never will be, and other parts of that same publication look quite embarrassing now. To be fair, parts of *On War* look fairly ridiculous now. However, it would be hard to know which parts will endure and which will not. To that extent, the contents of this book is not 'the answer'. It's a step along the way. Its contents will date.

We are all products of our time. The science of physics, and particularly the ideas of friction and the centre of mass (which is one translation of the word 'schwerpunkt') were exciting and novel to Clausewitz.[27] I was brought up with Keegan's *Face of Battle* and Dixon's *On the Psychology of Military Incompetence*. Archer Jones's magnum opus appeared when I was in my twenties, and I was introduced to systems theory in my early thirties. I was 29 years old when the Manoeuvrist Approach was first enunciated, although it came as little revelation. Surely everybody had read about the Israelis in the wars of 1956, 1967 and 1973, or the Wehrmacht in the Second World War? Some more senior officers, however, clearly found it a bit novel.

I have referred to the British Army throughout this book. There are many reasons for that, not least the availability of data and my familiarity with it. I have, however, worked with several other armies. I believe that we tend to underestimate the differences between armies because of their apparent similarities. Every army is unique and the differences can be profound. Nonetheless, I also believe that in general the contents of this book apply to most of them. Differences do exist. But that is what we should expect. If combat, and hence war, is fundamentally human, different human institutions will behave in different ways in war.

This book marries history with historical analysis (HA). As far as I am aware, Dupuy's works are the only others which take this approach. Some historians have criticized Dupuy's works on various grounds,[28] but I have cited virtually none of Dupuy's HA results. His work generally correlates quite well with that of British analysts, led by David Rowland. HA uses objective facts, identified by professional historians using rigorous historical methods. It is important to separate that from the opinion which historians, quite rightly, offer on the basis of their research: 'the acceptance of HA by historians has been divided into those that welcome this extra objective use of history and those that tend to suspect its trespass on the supremacy of historians' expert judgement'.[29] A key issue is to 'assemble a group of military historians who are both supportive of HA and capable of working with the appropriate self-discipline'.[30]

It's important to realize that HA rarely tells us anything we don't already know, at least to some extent. Aspects such as the actual effectiveness of shock and surprise, or the impact of control of the air on land operations at the campaign level, have long been surmised or vaguely anticipated. What HA seems to do is to put those things into proper focus: like focusing a pair of binoculars. Once one has done that, one can be surprised how much things which one vaguely knew were there really stand out. HA has allowed us to see what is actually important and to what extent. We should be very grateful to the few patient practitioners who work in the area.

The marriage of HA with history and extant military doctrine introduces another point. Mature sciences are just that: useful work can continue, and occasionally breakthroughs occur; but really groundbreaking, surprising advances are rare.[31] They tend to occur at the boundaries between existing disciplines. Thus, for example, the marriage of physics and medicine has allowed great advances in diagnostics and treatment of conditions such as cancer. So we should not be afraid to look to the boundaries of existing military thought to make further developments.

What we have found is typical of the emergence of a new paradigm in science: the discovery of anomaly; the formulation of a new approach by a relatively junior practitioner; a better description of extant phenomena (including the resolution of paradox); and the finding of new and hitherto unsuspected phenomena which appear useful. When Newton postulated the concept of gravity, everybody knew that dropped items fall to the ground. They knew that the sun appears to rise, pass across the sky and set. Some knew of the elliptical movement of heavenly bodies in space. Newton suggested that such behaviour was due to the mutual attraction between those bodies. He was able to quantify such behaviour, and be predictive.

Today, everybody knows that wars are waged between armed bodies. There is considerable anecdotal received wisdom that human aspects are the most important in war. Keegan's and Dixon's books explored those aspects in some detail. The behavioural human sciences are now broadly accessible to the interested general reader. This book has begun to show how what we observe in combat is basically due to human behaviour in a complex, lethal, dynamic environment. That behaviour is not yet quantifiable nor predictable. We can, however, at least make an initial causal description. War, and particularly combat, are essentially human activities. The way ahead is to explore the consequences of that behaviour, in that environment, in a properly structured manner. It lies largely in an exploration of the human behavioural sciences: psychology, the study of mental processes (and particularly applied psychology, the study of human interaction); sociology, the study of human institutions; and anthropology, the study of human culture. All of which leads one to conclude that a revolution in military thought may not be quite imminent.

Combat is ineluctably dominated by human behaviour, and the major condition for defeat is psychological: that the enemy believes he is beaten. Therefore armies should not just fight. They should manipulate the use, and the threat of the use, of violence. That being so, we will not witness a revolution in military affairs until armies become expert not just in the manipulation of the use of violence, but also in the manipulation *of the threat of the use of* violence. That would be a huge challenge for any Western army. Many insurgent or terrorist organizations seem to be much better at manipulating the threat of the use of violence than the security forces which they face.

So much for the theory; what about the practitioners? Our logic suggests that people who read this book will tend to fall into four overlapping groups. The first will be those who tend to pedantic exception. Little of combat is categoric, and much of our discussion consists of generalizations. It accepts exceptions: that is a strength, not a weakness, but pedants might not accept that. And, of course, a well-presented exception can ruin the acceptance of a broadly useful general case.

The second group will be those who reject the new paradigm. To them, combat 'isn't like that'. That is to be expected: it happens at the introduction of most novel paradigms.[32] If the novel paradigm is successful, their numbers will fall away, but possibly never disappear. Those remaining would be the military equivalent of the 'Flat Earth' community. Their numbers would stand in inverse proportion to the success of the new paradigm.

The third group will struggle to understand the new paradigm and its implications. There will be a surprisingly large number of them. Hopefully, that number will grow at the expense of the previous groups. I have been to hundreds of meetings in which most of the participants broadly agreed with each other over most issues. However, they typically spent a long time discussing them, because of the need to explore and internalize them and their implications. Much of the material discussed here will only be summarized in military publications, which would exacerbate the problem. Besides, this book is far from being a complete exposé of a revised paradigm and all its implications.

The fourth group would say that what is presented is useful and probably valid, but doesn't tell us anything we don't already know. This happened me once at a prestigious occasion. A retired senior officer made exactly that point. He may have overlooked that, although it may be factually correct, no one appears to have assembled all the material in one place before. If Vauban's only contribution to siegecraft was codification, then that was a most impressive contribution. Nevertheless, such remarks are to be hoped for. They reveal that the revised paradigm is actually being accepted.[33]

Thus the logical, personal level. That does not necessarily lead to the substantive, institutional change implied. At that point one must express doubt. The British Army is broadly successful, and perceives little need for change. Past disasters like the Fall of France or the *Kaiserschlacht* are insufficiently salient in its memory to prompt cultural change. An army's commanders may impose change, but the succession of posts and cultural persistence means that such change tends to be strongly resisted, and perhaps undone a few years later. The same will apply in any army.

Much remains to be done, and it requires the application of the behavioural human sciences to the battlefield. That in turn will require the development of a common language, sensible discourse and greater rigour. Operational research and historical analysis should be exploited more fully. In particular, operational research methods should be adapted to simulate human effects far more accurately than is currently the case.

We have largely discussed ground combat, and by implication major land-combat operations. The work should be extended into air–land operations, and into counter-insurgency and peace support operations. The will and cohesion of the protagonists in an insurgency or a fragile peace seem sensible targets to study and influence. The application of shock and surprise may be sensible, but only on those occasions when violence is applied. Thus in such conflicts much of the contents of this book may be interesting but irrelevant. It may be that much of it is inapplicable, and further theories are needed. So be it. Let us recognize the limitations of the present work, rather than try to make it fit where it patently doesn't.

In the last few weeks of 1899 the British Army, which had recently won a crushing victory over the Mahdist army in the Sudan, were given a nasty series of shocks by Boer forces in South Africa. The greatest risk to the army of a developed Western nation is complacency. To win, one only has to be better than the other guy, and Western armies generally think that they are. Unfortunately we don't know how good 'better' actually is. Armies should strive to be as good as they possibly can, at the things that really matter. An acquaintance of mine was a senior staff officer during the fighting phase of the Iraq War of 2003. His conclusions formed the final words of the army's analysis of those operations: 'We can do better.'[34]

This book is dedicated to the apocryphal British soldier, Tommy Atkins. Wellington once described his soldiery as 'the scum of the Earth, enlisted for drink'. Importantly, he added that it was wonderful 'that we should have made them the fine fellows that they are'.[35] In my experience the Duke's second comment still applies. A decade ago a very wise infantry brigadier suggested to his officers that, man for man, the British soldier was probably amongst the best in the world; but that 'he is as good, or as bad, as we allow him to be'.

How true. I'm sure that much the same applies in every army. War has a human face.

Notes

Notes to Introduction

1   Smith, R., p. 1.
2   Gray, p. 33.
3   Thucydides, in Gray, p. 30
4   Lind.
5   Luttwak.
6   Van Creveld (1985) and (1991).

Notes to Chapter 1: Art or Science?

1   Bronowski, p. 162.
2   *The Concise Oxford Dictionary*, p. 1310.
3   Ibid., p. 1311.
4   Keegan (1993), pp. 91ff.
5   *Design for Military Operations*, p. 4–3.
6   Irving, p. 9.
7   Howard, p. 33.
8   Gat (2001), pp. 253ff.
9   Trythall, p. 31.
10   Potts, pp. 10–15.
11   Demchak, p. 97.
12   Department of Health: www.doh.gov.uk/stats/doctors, accessed 19 May 2001.
13   For example, the civil engineering department of the University of Sheffield employed 25 academic and ten post-doctoral research staff: www.shef.ac.uk/civil/lings1/html, accessed 21 May 2001.
14   Gray, p. 227.
15   Ibid., p. 142.
16   Storr (2005), *passim*.
17   Lind, pp. 4–5.
18   *The Staff Officer's Handbook*, p. 3–1–1.
19   Kiszely, pp. 36–40.
20   Popper, p. 27.
21   Sims, p. 144.
22   Spick, pp. 147ff.
23   Spick, *passim*.
24   *Army Doctrine Publication (ADP) Command*, para. 0314.
25   Hampshire, p. 32.
26   *ADP Operations*, paras. 0227 and 0502.
27   Ibid., para. 0502.
28   *ADP Command*, para. 0103.

29   Ibid., para. 0106.
30   Pigeau and McCann pp. 475–90.
31   *Formation Tactics*, para. 0227.
32   McCann and Pigeau, loc, cit.
33   *Fighting Future Wars*, pp. 2–14.
34   Professor Richard Holmes, personal communication.
35   Dupuy (1977), p. 178.
36   *UK Doctrine for Joint and Multinational Operations*, Ch. 3.
37   This is not true. A principle is defined as a general law or a law of nature (*Concise Oxford Dictionary*, p. 880). It is difficult to consider laws of nature as having exceptions.
38   'A principle should admit of no exceptions. A principle is of a wider range of application than a rule, which is a more specific guide in a particular field.' (J.J. Davies, p. 26).
39   Exercise Sea Wall was a two-sided, force-on-force simulator exercise conducted by Headquarters Infantry in Cyprus in November/December 1996.
40   Gray, pp. 134, 142.
41   Clausewitz, p. 203.
42   Luttwak, pp. 94, 120 and *passim*.
43   Ibid., p. 62.
44   *Counterinsurgency*, pp. 1–22ff.
45   Hollis, p. 51.
46   Ibid.
47   Kuhn, p. 163.
48   Clausewitz, Book 2, Ch 3.
49   Hollis, loc. cit.
50   Professor Gary Sheffield, personal communication.
51   Kuhn, pp. 11–13.
52   Ibid., pp. 16–18.
53   Ibid., Ch 5, *passim*.
54   They were not common in mid-nineteenth-century military thought. Dupuy (1977), p. 134.
55   Kuhn, pp. 20 and 62.
56   For example, Archer Jones's characterization of four basic troop types and the pervading relationships between them. Jones, *passim*.
57   Popkin and Stroll, p. 231.
58   Hampshire, p. 143.
59   Kuhn, pp. 15–16.
60   White, *passim*.
61   Kuhn, loc. cit.
62   Ibid.
63   Colonel (now Brigadier) David Potts, personal communication.

Notes to Chapter 2: *Developing an Approach*

1   Popkin and Stroll, p. 133.
2   Ibid., p. 12.
3   Clausewitz, p. 202.
4   Shy, John. *Jomini*, in Paret (ed.) Ch. 6.
5   Keegan (1998), p. 342.
6   Ascribed to Hume in Bradley, p. 44.
7   Continuity, succession and the inference of causality were first enunciated by David Hume. Popkin and Stroll, pp. 254–60.

8 Ernst Mach, quoted in Bradley, pp. 11–12.
9 A Consequence of Heisenberg's Uncertainty Principle, Hawkins, p. 54.
10 Williams, G.P., p. 211.
11 Bradley, p. 139.
12 Carver (1978), p. 16.
13 Keegan (1998), p. 194.
14 Examples in the field of organizational psychology include Herzberg's *Motivation-Hygiene Theory* (1966), McGregor's *Theory X and Theory Y* (1960) and Simon's *Decision-making and Organizational Design* (1960). Pugh (ed.), Chs 11, 16 and 17.
15 Gat (2001), p. 255.
16 Keegan (1976).
17 Davies, J., p. 26.
18 Popkin and Stroll, pp. 234–5.
19 Gray, p. 88.
20 Popkin and Stroll, p. 268.
21 Frank, P. *The Laws of Causality and their Limits*, in Popper, p. 279.
22 Hampshire, p. 32.
23 I served for two years at the Defence Evaluation and Research Agency, liaising between scientists and soldiers. I have seen both sides of the discussion.
24 Popper, p. 278. Italics in original
25 Ibid., p. 281. Italics in original.
26 Demchak, p. 138.
27 Popkin and Stroll, p. 320.
28 Bradley, p. 107.
29 Townshend (ed.), p. 82. Although written in relation to colonial warfare, the text suggests that this process occurs 'as in all warfare'.
30 Fraser, p. 562.
31 Patton, pp. 403ff.
32 Von Manstein, *passim*.
33 Griffith (1994), p. 149.
34 Ibid., pp. 155–7.
35 Darryl Jaya-Ratnam, DERA Fort Halstead, personal communication. See also Jaya-Ratnam, *passim*.
36 *The Times*, 7 November 1999, p. 16.
37 During the 1990s the British Army's planning figure was 20 per cent of all casualties killed and missing. *Staff Officer's Handbook*, p. 3–41–1.
38 Major i. G. Bruno Paulus and Major i. G. Dirk Brodersen, separately, personal communication. Both are now colonels.
39 Popkin and Stroll, p. 343.
40 Paret (ed.), p. 186.
41 Dupuy observed 'typical' French concern for consistency between military theory and practices. Dupuy (1977), p. 113.
42 Geert Hofstede compared the differences between the French and German academic intellectual traditions and the British. see Hofstede, *passim*.
43 Popper, p. 280.
44 Howard, p. 32.
45 Popkin and Stroll, pp. 320ff.
46 Howard, Ch 3 Section VII.
47 Kuhn, p. 55.
48 Ibid., p. 97.
49 Paret, p. 172ff.
50 Kuhn, pp. 66ff.

51   Popper, pp. 49–50.
52   Ibid., pp. 99ff.
53   Doctor Heidi Doughty, personal communication.
54   T.E. Lawrence, in Dixon (1997), p. 157.
55   Clausewitz, p. 164.
56   Ibid., p. 133.
57   Fraser, p. 561.
58   Gat (2001), p. 205.
59   Gray, p. 45.
60   Nosworthy, p. 120.

Notes to Chapter 3: The Nature of Combat

1   Van Creveld, in Townshend (ed.), pp. 306ff.
2   This passage concentrates on a criticism of the doctrine of the British Army. Colleagues in other Western armies have acknowledged similar failings in their own armies.
3   *ADP Operations*, para. 0223. The paragraph which discusses the importance of deciding and acting faster than the enemy also suggests that superior tempo can be achieved 'by speeding up or slowing down.'
4   *Staff Officers' Handbook*, p. 3-1-1.
5   Kuhn, p. 149.
6   Dr Richard Holmes, presentation to Army Command and Staff Course, 28 December 1994.
7   Gat (2001), p. 183.
8   Williams, G.P., p. 13.
9   Ibid., p. 234.
10   Czerwinski, pp. 117–18.
11   Ibid., pp. 118–19.
12   Luttwak and Horowitz, p. 173.
13   Nosworthy, p. 154.
14   In which the power is almost always higher than two.
15   Von Bertalanffy, in Capra, pp. 43–50.
16   Lecture to Division One of the Army Command and Staff Course, March 1993.
17   A theory must contain a set of axioms that are free from contradiction, independent, necessary and sufficient. 'Necessary and sufficient' is a common scientific shorthand for a set of conditions to be valid. See Popper, p. 71.
18   The Yom Kippur and Indo-Pakistan wars; the Falklands conflict; the Soviet incursion in Afghanistan and the Iran–Iraq war.
19   Cordesman and Wagner, Vol. 1, pp. 8–13.
20   Zetterling and Frankson, pp. 145–8.
21   Howard, p. 30.
22   Ellis and Cox, p. 6.
23   Dupuy (1977), p. 151.
24   Fraser, p. 562.
25   Lieutenant-General Paul van Riper, Commanding General of the US Marine Corps Combat Development Center, prepared testimony to the House of Representatives National Security Committee, March 1997.
26   Rowland *et al.*, p. 4.
27   Ibid., p. 42.
28   Ibid., p. 29.
29   Townshend (ed.), p. 178.

30   Bar-On, pictures to Ch. 8.
31   Williams, G.P., p. 86.
32   Clausewitz, pp. 311–15.
33   Williams, J., p. 175.
34   Ellis and Cox, p. 270.
35   Bar-On, *passim.*
36   *ADP Operations*, Ch. 2.
37   Ferdinand Foch, *Principles of War*, p. 286, in Gat (2001), p. 402.
38   Dupuy (1977), p. 178.
39   Spartacus (1991), pp. 48–52.
40   Nosworthy, p. 27.
41   *ADP Operations*, paras. . 0206–10.
42   *In Praise of Attrition*, inaugural lecture by Christopher Bellamy as Professor of Military Sciences and Doctrine, Cranfield University, 14 June 2001.
43   Keegan (1998), pp. 305–6.
44   Nosworthy, p. 455.
45   I first heard it expounded by Brigadier Gage Williams (J.G. Williams), who had recently served as an instructor at the US Army's School of Advanced Military Studies, in 1991 or 1992. Discussion with the Brigadier and elsewhere failed to reveal any real provenance for the expression. Hence the attribution given here.
46   Delbruck (1990b), pp. 87–9.
47   Paret (ed.), pp. 105–13.
48   Strachan, pp. 23ff.
49   Dupuy (1977), p. 214.
50   Gat (1997), pp. 160–72.
51   Ibid., p. 172.
52   Army List, 1940.
53   Seaton, pp. 64–7.
54   Jones, pp. 603–8.
55   Ibid.
56   Welch, p. 294.
57   Gray, pp. 23–4.
58   Dupuy (1978), p. 601.
59   *Operations in Iraq*, p. 3–1.
60   Gat (2001), p. 193.
61   Kuhn, p. 85.
62   Hawkins, pp. 29–32.
63   Marmont, *The Spirit of Military Institutions*, p. 49, in Nosworthy, p. 24.

Notes to Chapter 4: *Tools and Models*

1   D'Este (2000), p. 287.
2   Gat (2001), p. 146.
3   Clausewitz, p. 122–38.
4   Howard, p. 35.
5   *ADP Operations*, para. 0210.
6   Ibid.
7   Von Bertalanffy, in Capra, pp. 43–50.
8   Jones, pp. 294–308.
9   Demchak, pp. 18–25.

10   Van Creveld (1983), p. 121.
11   Erskin, *passim.*
12   Van Creveld (1991), p. 121.
13   Jones, p. 494.
14   Ibid., pp. 111–12.
15   Verbruggen, pp. 111–203.
16   Bennett, pp. 1–18.
17   Jones, pp. 36–8.
18   Ibid., p. 566.
19   Ibid., pp. 36–8.
20   Ibid., p. 612.
21   Mackay-Dick, *passim.*
22   Whitaker *et al.*, pp. 10–11.
23   Kearney and Cohen, p. 106.
24   Ibid., p. 98.
25   Ibid., pp. 69–71.
26   Ibid., p. 113.
27   Whitaker *et al.*, pp. 227–8.
28   Ibid., p. 294.
29   Ellis (1993) pp. 184–5. Ellis lists only 23 destroyed, but one of those was counted twice.
30   Whitaker *et al.*, pp. 227–8 Whitaker lists 65,000 wounded, but presumably a proportion of the wounded were among the prisoners. Even if not, at least 70,000 able-bodied men escaped the Falaise Pocket.
31   D'Este (2000), p. 122.
32   Feuchtinger, in Isby (ed.), p. 116.
33   Carell, p. 29.
34   Middlebrook, p. 322.
35   The issue of air combat kill reporting is treated extensively in Morgan and Seibel.
36   Zetterling and Frankson, *passim.*
37   Larionov *et al.*, p. 516.
38   *Battles Hitler Lost*, p. 7.
39   Ellis (1993), pp. 129–39.
40   Ibid.
41   Ibid.
42   Keegan (1998), pp. 300–5.
43   Haythornwaite, p. 32.
44   Gudmundsson, pp. 66ff.
45   Stanton, *passim.*
46   Farrar-Hockley, p. 37.
47   Hutchinson, pp. 290 and 299.
48   Von Mellenthin, p. 65 and *passim.*
49   Jones, p. 641.
50   Nosworthy, Ch. 17.
51   Rooney *et al.*, pp. 13–14.
52   Van Creveld (1985), p. 270.
53   Bryan Stewart, Centre for Defence Analysis, personal communication.
54   Duffy (1987), p. 215.
55   *The Instructor's Handbook on Fieldcraft and Battle Drill*, sect. 48.
56   Verbruggen, p. 151.
57   Holmes (ed.), p. 318.
58   Where medieval infantry acquired operational effectiveness, confidence and social cohesion were

major factors: see Verbruggen, p. 179.

59 Delbruck (1990a), p. 548, n. 1.
60 *The Concise Oxford Dictionary of Quotations.*
61 Adan, p. 366.
62 D'Este (2000), p. 121.
63 Ibid. pp. 271, 274.
64 Delaforce (1994), pp. 12 and 66.
65 D'Este (2000), p. 263.
66 Ellis (1993), p. 125.
67 Ibid., pp. 132–3.
68 Delaforce (1997), pp. 43 and 113.
69 For example, the Adjutant of the 5th Black Watch, Captain Bill Bradford, escaped capture and subsequently commanded the new 5th Black Watch from Normandy to the Baltic. Lieutnant-Colonel R.J.K. Bradford, personal communication.
70 Similarly Captain D.B. Lang subsequently commanded the 5th Camerons in the reformed Division from July 1944. Obituary, *The Times*, 10 April, 2001.
71 Dupuy (1994), p. 395.
72 Ibid.
73 Ibid., p. 398.
74 *US Army Order of Battle WW2. European Theatre, Divisions*, pp. 573–4.
75 *The Procurement and Training of Ground Combat Troops*, 1948, pp. 428–93.
76 Dupuy (1977), p. 100.
77 Harrison Place, pp. 41–3.
78 And also in my experience. In British Regular infantry battalions, routine low-level training up to company level does often take a 'back seat' to barrack routine.
79 Van Creveld (1983), p. 123.
80 D'Este (2000), p. 272.
81 Lindsay (1981), p. 43.
82 Reynolds, D., pp. 372–3.
83 Ellis (1993), pp. 115–16.
84 Colville, p. 266.
85 O'Ballance, p. 245.
86 Keegan (1998), pp. 355–6.
87 Ibid.

Notes to Chapter 5: Shock and Surprise

1 Britain's *Glossary of Joint and Multinational Terms and Definitions* has only a definition for 'Suppressive Fire'.
2 Storr (2000), p. 3.
3 *The Effects of Shock and Surprise on the Land Battle, passim.*
4 Ibid.
5 Rowland (2006), p. 175.
6 Rooney *et al.*, p. 14.
7 Rooney (1997), *passim*.
8 Dupuy (1977), p. 256.
9 Van Creveld (1983), p. 6 and *passim*.
10 English, p. 92.
11 Storr (2002), p. 143.
12 David Rowland, personal communication.

13   *The Effects of Shock and Surprise on the Land Battle*, Table 2.
14   Foreman, p. 200.
15   Jaya-Ratnam, *passim.*
16   Such as the Russian use of armour in the first battle of Grozny, or Israeli tactics in Suez City in 1973.
17   Rowland (1991), pp. 156–7.
18   Kuhn, pp. 152–3.
19   O'Ballance, p. 246.
20   Guderian (1976), pp. 102 and 109.
21   *AFM Formation Tactics*, para. 0226.
22   *ADP Operations*, para. 0214.
23   Stephen Ashford, Senior Analyst at DSTL, personal communication.
24   Griffith (1994), p. 76.
25   Ibid., pp. 84–6.
26   Stanton, pp. 365–6.
27   Gudmundsson, p. 168.
28   Keegan (1998), p. 308.
29   *ADP Operations*, para. 0234.
30   *Der Angriff in Stellungskrieg* of 26 January, 1918 (the Hindenburg Regulations), in Gudmundsson, p. 150.
31   Balck, p. 357.
32   Beca, p. 56.
33   Gudmundsson, p. 85.
34   Liddell Hart, pp. 38–44.
35   Van der Bijl, pp. 170–88 and *passim.*
36   Ibid., p. 140.
37   Harrison Place, pp. 1–3.
38   Ibid., p. 68.
39   Field Marshall Erwin Rommel, in English, p. 100.
40   *Fighting Future Wars*, p. 12.
41   Identifying and satisfying aims: Conformity of the Goal. The nature of the means: Preservation of Combat Effectiveness. The application of the means: Speed and Shock, Concentration of Effort, Surprise and Security, Combat Activeness and Coordination of Forces (from Isby, p. 13).
42   *ADP Operations* para. 0231.
43   Keegan, (1998), p. 302.
44   Ian Gardner, DERA Fort Halstead, private communication.
45   My personal observation.
46   Swinson, p. 13.
47   Ibid., p. 43.
48   Spiller, pp. 10–11.
49   *Operations in Iraq*, para. 321.
50   *ADP Operations*, para. 0534.
51   *AFM Formation Tactics*, para. 0110.
52   Fuller (1932), pp. 212–6.
53   Major-General (retired) Keith Spacie, personal communication.
54   *ADP Operations* para. 0534.
55   Ibid.
56   Whitaker *et al.*, pp. 227–8 and *passim.*
57   D'Este (2000), pp. 162–3 and 232.
58   Carrell, pp. 120–25.
59   21st Panzer Division.

60 Rowland (1991), pp. 149–62.
61 Storr (2000), p. 7.
62 Lynam, pp. 32–43.
63 Rowland, (1991), pp. 543–53.
64 Lynam, *loc. cit.*
65 *AFM Operations in Built-Up Areas, passim.*
66 Storr (1994).
67 The Chindit Memorial, Westminster Embankment, London.
68 Van Creveld, in Townshend (ed.), pp. 298–314.
69 Gray, p. 116.
70 Bellamy (1990), pp. 236–45.
71 Van Creveld (1985), p. 269.
72 Demchak, p. 82, n. 19.

Notes to Chapter 6: *Tactics and Organizations*

1 Guderian (1992), p. 183.
2 English and Gudmundsson, pp. 48–52.
3 Gat (2001), pp. 162–6.
4 *The Procurement and Training of Ground Combat Troops*, pp. 448–50.
5 *The Instructor's Handbook of Fieldcraft and Battle Drill (Provisional), passim.*
6 Eastaway and Wyndham, pp. 73–8.
7 Patton, pp. 403ff.
8 *The Attack in Position Warfare*, 26 January, 1918, (the Hindenburg Regulations, in Gudmundsson, p. 150.
9 *Night Combat*, p. 36.
10 Van der Bijl, p. 171.
11 Whitaker *et al.*, pp. 55–6 and *passim.*
12 Ibid., pp. 127–8.
13 Exercise 'Sea Wall'.
14 Colonel (ret.) Yoav Hirsch, private communication.
15 *The Kirke Report*, section 13.
16 *Combat in Russian Forests and Swamps*, p. 25.
17 Beevor (1992), pp. 145–6.
18 E.g. *The Kirke Report, loc. cit.*
19 E.g. map at D'Este (2000), pp. 134–5.
20 Zetterling and Frankson, p. 96.
21 Luttwak and Horowitz, pp. 237–41.
22 *The Principles of Employment of Armor*, pp. 20–21.
23 Oman (1987), p. 260.
24 Patton, p. 413.
25 Griffith (1994), p. 98.
26 Perret, p. 65.
27 Gudmundsson, p. 151.
28 *Sixth Armored Division 1944–5*, p. 45.
29 Griffith, (1994), p. 99.
30 Patton, p. 263, quoting Kipling's 'If'.
31 Ellis (1993), p. 219.
32 Establishment Tables: II/100/3; II/104/3; III/181/2; II/261/2; III/290/1.
33 Van Creveld (1983), pp. 49–52.

34 The Second Panzer Division lost its panzer brigade HQ in August 1940, and its infantry brigade HQ on 30 September 1942. Strauss, pp. 240–41.
35 Ellis, (1993), pp. 18–19.
36 *Staff Officer's Handbook*, pp. 1-29-1ff.
37 Letter from Army Historical Branch, HB(A) 6/3 dated 1 Dec 98.
38 Roberts, pp. 183, 190–91 and *passim*.
39 Patton, p. 400.
40 *Staff Officer's Handbook*, p. 3-1-1.
41 *Army Supply and Transport*, Vol. I, p. 59.
42 Lieutenant-Colonel (retired) Lou Mahanty, personal communication.
43 *Army Supply and Transport*, Volume II, pp. 313–19.
44 *Staff Officer's Handbook*, pp. 1-120-1 and 121-1.
45 Dupuy (1977), pp. 177–8.
46 *Army Medical Services – Administration*. Vol. 1, p. 465.
47 Van Creveld (1991), p. 121.
48 Rooney *et al.*, pp. 13–16.
49 Maps Series M709, Editions 3-DMA, US National Imaging and Mapping Agency, Sheets 3079/I, 3080/II–III, 3178/IV, 3179/III–IV.
50 Foss, pp. 10–124 and 326–99.
51 *Staff Officer's Handbook*, pp. 1-29-1. Author's calculations.
52 Marsden, p. iii.
53 Dupuy, in Anthony, pp. 650–59.
54 Van Creveld (1985), p. 144.
55 David Watson, King's College London, unpublished research. Watson conducted the research, under contract to DERA CDA, from primary sources (unit and formation histories in the PRO and archives).
56 HB(A) Letter 11/8/1 dated 17 November 2000.
57 The Division typically fought two to three separate and identifiable engagements each day. The forces which fought in each engagement are known down to sub-unit level.
58 Von Manstein, in von Manstein and Fuchs, pp. 389–408. Trans. Ghislaine Fluck, DERA CDA, with my assistance.
59 Ibid.
60 *US Army. Order of Battle, World War 2. European Theater, Divisions*, p. 567.
61 Van Creveld (1991), pp. 105–7.
62 Nafziger (2000), pp. 31–3.
63 Of which one was a parachute brigade in half tracks. Bar-On, pp. 33–41.
64 I was the military adviser to the trial.
65 *Staff Officer's Handbook*, pp. 1–61–1.
66 Army Groups, Royal Artillery. Hastings (1984), pp. 345–6.
67 Isby, pp. 121, 161–2.
68 Von Mellenthin, p. 65 and *passim*.
69 *The Organization of Ground Combat Troops*, pp. 423–6.
70 Nafziger (1998), p. 28.
71 Rowland (2006), p. 122.
72 Guderian (1976), p. 43.
73 Garbert, pp. 63–4.
74 Ibid., p. 69.
75 Manstein and Fuchs, p. 400, n. 2.
76 Nafziger, (1998), p. 320.
77 *Staff Officer's Handbook*, loc. cit.
78 Establishment Table II/100/3.

79   *Staff Officer's Handbook*, p. 1-77-1.
80   Illustrative figures in *Staff Officer's Handbook* give 88 per cent ammunition by weight; the remainder mostly fuel, and 1–2 per cent rations and technical stores. Author's calculations from *Staff Officer's Handbook* pp. 3-37-1–5.
81   Establishment Table II/261/2.
82   Manpower and vehicle numbers were calculated using company sizes in the *Staff Officer's Handbook*. Totals were probably accurate to within 1,000 men and 100 vehicles overall.

Notes to Chapter 7:   Commanding the Battle

1   Pugh (ed.), pp. 30–42 and 107.
2   Demchak, pp. 18–9 and *passim*.
3   March *et al.*, *passim*.
4   Ibid.
5   'Seek the information you need for your own decisions, not other peoples. Seek it only when you need it. This creates a degree of calm.' Major-General Rupert Smith, lecture to the army Staff College, 29 June 1994.
6   Klein, p. 93.
7   Czerwinski, p. 235.
8   Kiszely (1999), *passim*.
9   Isby, p. 48.
10   Farrar-Hockley, pp. 16–17.
11   Taylor, *passim*.
12   Roberts, p. 201.
13   Bauer, pp. 132–6.
14   Piekaliewicz, p. 28.
15   British Army Command and Staff Course 28, Visit to HQ 3 (UK) Division, May 1995. Author's recollection.
16   *Staff Organizations and Operations*, pp. 6–7.
17   Patton, p. 408. Emphasis in original.
18   Rouse, pp. 42–5.
19   Ibid.
20   The use of inverted commas here is deliberate. The terms given are intended to be relatively vague and conceptual.
21   Klein, p. 275.
22   Such time delays are quantified and discussed in *The Contribution of Manned Armoured Reconnaissance to the Information Gathering Assets of 1 (BR) Corps – Exercise White Ermine*, *passim*.
23   Army Historical Branch letter HB(A) 6/3 dated 1 December 1998 and Major Grant Hume, personal communication.
24   The issue of organizational complexity and decision-making is treated at length in Czerwinski, pp. 140–44 and Demchak, pp. 27–36 and pp. 90–96.
25   *Operations in Iraq*, paras. 419–21.
26   Dermot Rooney, personal communication.
27   HQ 20 Armd Bde 20/G/001, dated 13 May 74.
28   *Operations in Iraq, loc. cit.*
29   Dermot Rooney, personal communication.
30   HB(A) Letter 6/3, dated 1 December 1998.
31   *The Twentieth-Century Variation of C2 Over Time*, *passim*.
32   Antal, pp. 12–16.
33   Van Creveld (1985), p. 40.

34 Czerwinski, p. 3.
35 There is a strong suggestion that the Schlieffen Plan, contrary to the spirit of Moltke the Elder, failed because Schlieffen attempted to plan the whole operation a priori. Paret (ed.), p. 323.
36 Rooney *et al.*, p. 16.
37 'It is the mark of an educated mind to rest satisfied with the degree of precision that the nature of the subject admits, and not seek exactness when only an approximation is possible.' Aristotle, *Nicomachean Ethics*, quoted in Czerwinski, p. 41.
38 Klein, p. 232.
39 Patton, p. 357.
40 Guderian (1976), Appendix VI.
41 Jacobsen gives all operations orders written at corps level and above (up to OKH) in the French campaign of 1940.
42 Storr (2001b).
43 Patton, p. 357.
44 Colonel Nick Metcalfe, private communication.
45 Klein, p. 93.
46 Czerwinski, p. 152.
47 George Brander, DERA Centre for Human Sciences, personal communication.
48 Klein, pp. 19–26.
49 Ibid., p. 52.
50 Rooney *et al.*, p. 4.
51 Pennington and Hasty, quoted in Klein, Ch. 11.
52 Klein, p. 103.
53 Ibid., pp. 97–101.
54 Rouse, pp. 42–5.
55 Interview in *Command of Armour, passim.*
56 *ISTAR Integration Warfighting Experiments 2002*, p. 30.
57 Andy Belyavin, DERA Centre for Human Sciences, personal communication.

Notes to Chapter 8: The Soul of an Army

1   Spartacus (1991), pp. 48–52.
2   Fraser, p. 8.
3   Holmes (ed.), pp. 598–9.
4   Gordon, p. 182.
5   Jackson, Appendix 3.
6   Army List, April 1940.
7   Sheffield and Todman (eds), p. 152 and *passim.*
8   Lieutenant-Colonel (now Brigadier) Geoff Sheldon, personal communication.
9   Corum, pp. 85–101.
10 Only 84 of the 2,344 were born after 1900 (and so would have been 45 or younger in 1945). Stumpf, p. 289 (author's translation).
11 Westphal, pp. 63–4. I have seen a partial list of names which indicates that roughly one in four German lieutenant- and major-generals died in service during the Second World War.
12 Major-General Tom Rennie of 51st Highland Division, killed by shellfire at the Rhine crossings. D'Este (2000), p. 276.
13 Dupuy (1977), p. 4.
14 Van Creveld (1983), p. 6.
15 Beevor (1998), p. 204.
16 Senich, pp. 16ff.

17   Reynolds, pp. 246–7.
18   Williamson, pp. 191–2.
19   Sims, pp. 30, 132 and Appendix III.
20   Ibid., p. 134.
21   Roscoe, p. 137. Submarines are known as 'ships', not 'boats'.
22   Author's estimate. The USN laid down 315 submarines in the war. If we assume that numbers are pro
     rata to (submarine) losses, about 290 served in the Pacific.
23   Tarrant, p. 146.
24   Author's estimate. The Kriegsmarine built 1,141 U-boats (ibid., p. 280).
25   Senich, pp. 16ff.
26   Rohwner, p. 98.
27   Tarrant, pp. 152–9.
28   Doctor George Brander, personal communication.
29   Williamson, pp. 117–27.
30   Senich, pp. 16ff.
31   Most were members of the Small-arms School Corps.
32   Rowland (2006), pp. 40, 116–17 and 142.
33   Morgan and Siebel, p. 166.
34   Sims, p. 31.
35   Senich, pp. 16ff.
36   Storr (2000), p. 11.
37   D'Este, (2000), pp. 178–83.
38   Sometimes it took personal intervention from the CO. *The Warrior Class*, pp. 27–36.
39   Beevor, (1998), p. 31.
40   Ellis (1993), pp. 241 and 259.
41   Holmes (ed.), p. 482.
42   Duffy, pp. 310–12.
43   Obituary, *The Times*, 25 June, 2001.
44   Obituary, *The Times*, 6 June, 2001.
45   Obituary, *The Daily Telegraph*, 22 October, 2001.
46   Army List, spring 1973.
47   Major (retired) J. A. Storr, personal communication.
48   Major G.R. Mitchell enlisted in 1941 aged 19. He served until 1977, when he retired aged 55. Obituary,
     The Times, 11 October 2001.
49   De la Billière, p. 93.
50   *The Warrior Class*, pp. 27–36.
51   Army List 1995.
52   Lindsay (2000), p. 3.
53   Army List, February 1946.
54   Winton, *passim*.
55   Obituary, *The Daily Telegraph*, 22 October, 2001.
56   Holmes (ed.), pp. 598–9.
57   Ibid., p. 989–91.
58   Von Manstein, *passim*.
59   Fraser, *passim*.
60   Holmes (ed.), pp. 381–2.
61   Chaney pp. 21 and 146.
62   Colville, *passim*.
63   *Report of the Committee on the Lessons of the Great War*, p. 8.
64   Holmes (ed.), p. 366.
65   Colville, pp. 225 and 235.

66   Lewin, p. 2.
67   Holmes (ed.), pp. 843–4.
68   Caddick Adams, pp. 31–40.
69   Dupuy, (1977), pp. 212ff.
70   Colonel (ret.) Paul Lefever, personal communication.
71   Ashworth, pp. 86–9.
72   Carver (1988), Frontispiece.
73   *Command of Armour in World War II, passim.*
74   Major-General Jonathan Bailey, personal communication.
75   Major-General Mike Charlton-Weedy, personal communication. The General conducted a major
     statistical study of career patterns during the 1990s.

Notes to Chapter 9: Regulators and Ratcatchers

1   Arnold, pp. 30–32.
2   Kuhn, pp. 11–22.
3   Aidman and Schofield, *passim.*
4   Keirsey, pp. 12–20.
5   Assembled from Keirsey, Chs 3–6
6   Ibid.
7   Major Ian Byrne, staff of DERA CDA, personal communication. (Major Byrne is a qualified MBTI
     typologist.)
8   Professor Gary Sheffield, personal communication.
9   Dixon (1977), pp. 257–87.
10   Gordon, pp. 326, 383 and *passim.*
11   Doctor Ken Chaplin, principal psychologist to the DERA Centre for Human Sciences, personal
     communication.
12   Brewster Smith, pp. 159–63.
13   See also Dixon, p. 257.
14   Dixon, pp. 260 and 282.
15   Joanne Suddaby-Smith, personal communication.
16   *Command in Armour in World War II, passim.*
17   Doctor George Brander, personal communication.
18   Joanne Suddaby-Smith, *loc. cit.*
19   Colville, p. 2 and *passim.*
20   Blumenson, pp. 123–6, 241, 288.
21   Ken Chaplin, *loc. cit.*
22   Stuttaford, *passim.*
23   Sale, pp. 3–27, 1992.
24   My terminology. Sale calls this group 'HCSC', which is confusing because it is not true of all subjects
     in the group.
25   Sale, pp. 15–20.
26   Storr (2002), pp. 248–9.
27   Sale, pp. 18–19.
28   Joanne Suddaby-Smith, *loc. cit.*
29   Sale, p. 16.
30   Joanne Suddaby-Smith, *loc. cit.*
31   Sale, *loc. cit.*
32   Ibid.
33   Czerwinski, p. 155.

34 Lieutenant-Colonel Alan Gaw, senior dental officer, Dhekelia Garrison, personal communication.
35 Joanne Suddaby-Smith, *loc. cit.*
36 Barry, p. 59.
37 Brigadier Simon Caraffi, personal communication.
38 Dr Eugene Aidman, personal communication.
39 Major Ian Byrne, personal communication.
40 Myers and Kirby, pp. 16–17.
41 Hirsch and Kunnerow, p. 9.
42 Ibid., p. 13–25.
43 Major General Jonathan Bailey, personal communication.
44 Sale, pp. 11–12.
45 Dr Stephen Harland, personal communication.
46 Barry, personal communication.
47 Major General Christopher Elliot, personal communication.
48 Aidman and Schofield, pp. 1–6.
49 Graduates comprised a minority of officer entrants to Sandhurst in 1981. In 1995–96 they comprised 85 per cent.

Notes to Chapter 10: The Human Face of War

1 Luttwak, p. 20.
2 Cordesman and Wagner, p. 353.
3 Luttwak and Horowitz, p. 183.
4 Stanton, pp. 76–80 and *passim.*
5 Nicol, pp. 37–46.
6 Clausewitz, p. 166.
7 Van Creveld (1991), p. 108.
8 Patton, p. 413.
9 Delaforce (1994), pp. 27 and 115.
10 Griffith (1994), p. 149.
11 Hastings (1984), pp. 116–21.
12 A quick numerical comparison of the divisional memorials on Omaha Beach makes the point.
13 Whitaker *et al.*, p. 123.
14 Dupuy (1977), p. 237.
15 Van Creveld (1983), pp. 151–5.
16 Dupuy, (1977), p. 286.
17 Fraser, p. 273.
18 Spartacus (1997), pp. 90–91.
19 Dupuy, (1977), p. 24.
20 Bird, pp. 46–9.
21 *Operations in Iraq*, paras. 603–4.
22 Ibid., para. 446.
23 Gray, p. 115.
24 Ibid., pp. 98–102.
25 Ibid., p. 122.
26 Primarily from response to the writing of, and teaching of, *ADP Land Operations.*
27 Gat (2001), pp. 173–5.
28 Professor Gary Sheffield, personal communication.
29 Rowland (2006), p. 217.
30 Ibid., p. 218.

31 Kuhn, p. 38.
32 Ibid., p. 82.
33 Ibid., pp. 85 and 153.
34 *Operations in Iraq*, para. 609.
35 Elizabeth Longford, Wellington: The years of the Sword (London: Weidenfeld & Nicholson, 1969), in Holmes, p. 149.

Bibliography

Many of the works referred to have no acknowledged author, hence a simple alphabetical list by author would not be appropriate. Items are listed by sections as follows:

1 Books
2 British Military publications
3 US Army publications
4 Theses, papers and presentations
5 Articles and letters
6 Reports

If a work cited in the text has an acknowledged author, consult sections 1, 4, 5 and 6. If a short title is cited, consult Sections 2, 3 4, 5 and 6.

1: BOOKS

Adan, Major-General Avraham. *On the Banks of the Suez*, London: Arms & Armour Press, 1980.

Ashworth, Tony. *Trench Warfare 1914–18. The Live and Let Live System*, London: Pan Books, 1990.

Atteridge, A.H. *Famous Modern Battles*, London: Thomas Nelson, 1913.

Arnold, John *et al. Work Psychology. Understanding Human Behaviour in the Workplace*, London: Pitman, 1991.

Balck, Colonel, trans. Walter Kruger. *Tactics*, New Haven: Greenwood Press, 1977.

Bar-On, Colonel Mordechai (ed.). *The Israeli Defence Forces, The Six-Day War*, Slough: Foulsham, 1968.

Bauer, Cornelis, trans. D.R. Welsh. *The Battle of Arnhem. The Betrayal Myth Refuted*, London: Hodder & Stoughton, 1966.

Beca, Colonel, trans. Captain A. F. Custance. *A Study of the Development of Infantry Tactics*, London: Swan Sonnenschein, 1911.

Beer, Stafford. *The Heart of the Enterprise (The Managerial Cybernetics of Organization)*, Chichester: John Wiley, 1979.

Beevor, Anthony. *Crete. The Battle and the Resistance*, London: Penguin, 1992.

Beevor, Anthony. *Stalingrad*, London: Penguin, 1998.

Bellamy, Christopher. *The Evolution of Modern Land Warfare. Theory and Practice*, London: Routledge, 1990.

Bidwell, Shelford and Graham, Dominick. *Firepower: British Army Weapons and Theories of War 1904–45*, London: Allen & Unwin, 1982.

Bijl, Nicholas van der. *Nine Battles to Stanley*, Barnsley: Leo Cooper, 1999.

Billière, General Sir Peter de la. *Storm Command*, London: HarperCollins, 1992.

Blumenson, Martin and Stokesbury, James L. *Masters of the Art of Command*, New York: Da Capo Press, 1990.

Blumenson, Martin. *Patton – the Man Behind the Legend 1885–1945*, New York: William Morrow, 1988.

Bond, Brian. *The Pursuit of Victory*, Oxford: Oxford University Press, 1996.

Bradley, J. *Mach's Philosophy of Science*, London: Athlone Press, 1991.

Bronowski, Jacob. *The Ascent of Man*, London: Futura, 1981.

Capra, Fritjof. *The Web of Life: A New Synthesis of Mind and Matter*, London: HarperCollins, 1997.

Carell, Paul, trans. E. Osers. *Invasion – They're Coming!*, London: Transworld, 1963.

Carver, F.M. Lord. *The Apostles of Mobility*, London: Weidenfeld & Nicolson, 1979.

Carver, F. M. Lord. *Britain's Army in the 20th Century*, London: Pan Books, 1988.

Chaney Jr, Otto Preston. *Zhukov*, Norman, OK: University of Oklahoma Press, 1971.

Clausewitz, Carl von, ed. Anatol Rapoport. *On War*, London: Penguin, 1982.

Colville, J. R. *Man of Valour: The Life of F.M. Lord Gort VC GCB DSO MVO MC*, London: Collins, 1972.

Cordesman, Antony and Wagner, Abraham. *The Lessons of Modern War*, London: Mansell, 1990.

Creveld, Martin van. *The Military Lessons of the Yom Kippur War*, London: The Washington Papers, No. 24, Sage, 1975.

Creveld, Martin van. *Fighting Power. German and US Army Performance 1939–45*, London: Arms & Armour Press, 1983.

Creveld, Martin van. *Command in War*, Cambridge, MA: Harvard University Press, 1985.

Creveld, Martin van. *The Transformation of War*, New York: Macmillan, 1991.

Czerwinski, Tom. *Coping with the Bounds: Speculations on Non-Linearity in Military Affairs*, Washington, DC: Institute for National Strategic Studies, 1998.

Davies, D. R. and Shackleton, V. J. *Psychology and Work*, London: Methuen, 1975.

Davies, J. J. *On the Scientific Method*, Harlow: Longman, 1968.

Delaforce, Patrick. *Churchill's Desert Rats. From Normandy to the Baltic with the 7th Armoured Division*, Stroud: Alan Sutton, 1994.

Delaforce, Patrick. *Monty's Highlanders. 51st Highland Division in World War 2*. Brighton: Tom Donovan, 1997.

Delbruck, Hans, trans. Walter J. Renfroe Jr. *Medieval Warfare. History of the Art of War, Volume III*. Lincoln, NB: University of Nebraska Press, 1990a.

Delbruck, Hans, trans. Walter J. Renfroe Jr. *The Dawn of Modern Warfare. History of the Art of War, Volume IV*, Lincoln, NB: University of Nebraska Press, 1990b.

Demchak, Chris C. *Military Organizations, Complex Machines. Modernisation in the US Armed Services*, Ithaca, NY: Cornell University Press, 1991.

Dixon, Norman F. *On The Psychology of Military Incompetence*, London: Jonathan Cape, 1977.

Dixon, Norman F. *Our Own Worst Enemy*, London: Jonathan Cape, 1987.

Duffy, Christopher. *Siege Warfare. The Fortress in the Early Modern World 1494–1660*, London: Routledge & Kegan Paul, 1979.

Duffy, Christopher. *The Military Experience in the Age of Reason*, London: Routledge & Kegan Paul, 1987.

Dupuy, Trevor N., *A Genius for War. The German Army and General Staff, 1807–1945*, London: Macdonald & Janes, 1977.

Dupuy, Trevor N. *Elusive Victory. The Arab–Israeli Wars, 1947–74*, London: Macdonald & Janes, 1978.

Dupuy, Trevor N. *et al. Hitler's Last Gamble. The Battle of the Bulge, December 1944–January 1945*, Shrewsbury: Airlife Publishing, 1994.

Eastaway, R. and Wyndham, J. *Why Do Buses Come in Threes? The Hidden Mathematics of Everyday Life*, London: Robson Books, 1998.

Ellis, John. *The Sharp End of War. The Fighting Man in World War II*, Newton Abbot: David & Charles, 1980.

Ellis, John. *The World War Two Databook*, London: Aurum, 1993.

Ellis, John and Cox, Michael. *The World War One Databook*, London: Aurum, 2001.

English, John A. *A Perspective on Infantry*, New York: Praeger, 1981.

English, John A. and Gudmundsson, Bruce I. *On Infantry*, New York: Praeger, 1994.

Erickson, John. *The Road to Stalingrad* and *The Road to Berlin. Volumes One and Two of Stalin's War on Germany*, London: Grafton, 1985.

D'Este, Carlo. *Fatal Decision. Anzio and the Battle for Rome*, London: CollinsFontana, 1992.

D'Este, Carlo. *Decision in Normandy*, London: Robson Books, 2000.

Farrar-Hockley, Anthony. *Infantry Tactics*, London: Almark, 1976.

Foreman, D. *To Reason Why*, London: André Deutsch, 1991.

Forty, George and Duncan, John. *The Fall of France. Disaster in the West 1939–40*, Tunbridge Wells: Nutshell, 1990.

Foss, Christopher F. *Janes' Tank Recognition Guide*, Glasgow: HarperCollins, 1997.

Fraser, David. *Knights Cross. A Life of Field Marshal Erwin Rommel*, London: HarperCollins, 1993.

Freedman, Lawrence (ed.) *War*, London: HarperCollins, 1992.

Fuller, Major-General J.F.C. *Dragon's Teeth*, London: Constable, 1932.

Fuller, Major-General J.F.C. *Armoured Warfare*, London: Eyre & Spottiswoode, 1943.

Fuller, Major-General J.F.C. *The Decisive Battles of the Western World and their Impact on History. Volume One, 480 BC–1757*, ed. John Terraine, St Albans: Paladin, 1975.

Garbert, Dr Christopher R. *Seek, Strike and Destroy: US Army Tank Destroyer Doctrine in World War II*, Fort Leavenworth: Leavenworth Papers, No. 12, Combat Studies Institute, US Army Command and General Staff College, 1985.

Gardner, Howard. *Frames of Mind. The Theory of Multiple Intelligences*, London: Heinemann, 1983.

Gat, Azar. *A History of Military Thought. From the Enlightenment to the Cold War*, Oxford: Oxford University Press, 2001.

Geraghty, Tony. *The Irish War*, London: HarperCollins, 1998.

Gleick, James. *Chaos. Making a New Science*, London: Abacus, 1997.

Glen, Fredrick. *The Social Psychology of Organizations*, London: Methuen, 1975.

Gooch, J. and Perlmutter, A. (eds). *Military Deception and Strategic Surprise*, London: Frank Cass, 1984.

Goodwin, Jason. *Lords of the Horizons. A History of the Ottoman Empire*, London: Vintage, 1998.

Graham, Dominick. *Against Odds. Reflections on the Experiences of the British Army, 1914–45*, London: Macmillan, 1999.

Gordon, Andrew. *The Rules of the Game. Jutland and British Naval Command*, London: John Murray, 1996.

Griffith, Paddy. *Forward into Battle*, Swindon: The Crowood Press, 1981.

Griffith, Paddy. *Battle Tactics of the Western Front – The British Army's Art of Attack 1916–18*, London: York University Press, 1994.

Griffith, Paddy *et al. Wellington: Commander. The Iron Duke's Generalship*, Chichester: Anthony Bird, undated.

Guderian, General Heinz, trans. Constantine Fitzgibbon. *Panzer Leader*, London: Futura, 1976.

Guderian, Major-General Heinz, trans. C. Duffy. *Achtung – Panzer!*, London: Cassell, 1992.

Gudmundsson, Bruce I. *Stormtroop Tactics: Innovation in the German Army 1914–18*, New York: Praeger, 1989.

Harris, J.P. and Toase, F.N. (eds). *Armoured Warfare*, London: Batsford, 1991.

Harrison Place, Timothy. *Military Training in the British Army, 1940–44*, London: Frank Cass, 2000.

Hawkins, Stephen. *A Brief History of Time*, London: Guild Publishing, 1998.

Hampshire, Stuart. *The Age of Reason. The 17th Century Philosophers*, New York: Meridian, 1956.

Hastings, Max. *Overlord. D-Day and the Battle for Normandy 1944*, London: Michael Joseph, 1984.

Hastings, Max. *The Korean War*, London: Michael Joseph, 1987.

Haythornthwaite, Philip J. *The World War One Source Book*, London: Arms & Armour Press, 1992.

Herzog, Chaim. *The War of Atonement*, London: Weidenfeld & Nicolson, 1975.

Hirsch, Sandra Krebs and Kunnerow, Jean M. *Introduction to Type in Organizations*, Oxford: Oxford Psychological Press, 2000.

Hofstede, Geert. *Cultures and Organizations. Software of the Mind*, Maidenhead: McGraw-Hill, 1991.

Hollis, Martin. *Introduction to Philosophy*, Oxford: Basil Blackwell, 1985.

Holmes, Richard. *Redcoat. The British Soldier in the Age of Horse and Musket*, London: HarperCollins, 2001.

Holmes, Richard (ed.). *The Oxford Companion to Military History*, Oxford: Oxford University Press, 2001.

Howard, Michael. *Clausewitz*, Oxford: Oxford University Press, 1983.

Hutchinson DSO MC, Colonel G. S. *Machine Guns: Their History and Tactical Employment. A History of the Machine Gun Corps, 1916–22*, London: Macmillan, 1938.

Isby, David C. *Weapons and Tactics of the Soviet Army*, London: Jane's, 1988.

Inauer, Josef (ed.). *The Swiss Army 2000*, Frauenfeld: Huber, 2000.

Jacobsen, Hans-Adolf. *Documente der Westfeldzug*, Berlin: Musterschmidt, 1960.

Jary, Sidney. *18 Platoon*, Bristol: Sidney Jary, 1994.

Jackson, Robert. *The Fall of France. May–June 1940*. London: Weidenfeld & Nicolson, 1975.

Jomini, Baron Antoine Henri de. *The Art of War*, London: Greenhill Books, 1992.

Jones, Archer. *The Conduct of War in the Western World*, London: Harrap, 1998.

Jüngner, Ernst. *The Storm of Steel*, London: Constable, 1994.

Kearney, Thomas A. and Cohen, Eliot A. *Gulf War Airpower Survey: Summary Report*, Washington, DC: US Government Printing Office, 1993.

Keegan, John. *The Face of Battle*, London: Jonathan Cape, 1976.

Keegan, John. *Six Armies in Normandy. From D-Day to the Liberation of Paris, June 6th–August 25th, 1944*, London: Book Club Associates, 1982.

Keegan, John. *A History of Warfare*, London: Hutchinson, 1993.

Keegan, John. *The First World War*, London: Hutchinson, 1998.

Keeley, Kevin. *The Longest War*, Westport, CT: Lawrence Hill, 1988.

Keirsey, David. *Please Understand Me II: Temperament, Character and Intelligence*, Del Mar, CA: Prometheus Nemesis, 1998.

Klein, Gary. *Sources of Power: How People Make Decisions*, London: MIT Press, 1998.

Kuhn, Thomas. *The Nature of Scientific Revolutions*, Chicago, IL: University of Chicago Press, 1996.

Larionov, V. *et al.*, trans. William Biley. *World War II: Decisive Battles of the Soviet Army*, Moscow: Progress Publishers, 1984.

Larson, Robert H. *The British Army and the Theory of Armored Warfare, 1918–40*, London: Associated University Press, 1984.

Leigh, Andrew. *Decisions, Decisions*, Aldershot: Gower, 1983.

Lewin, Ronald. *Slim the Standard Bearer. A Biography of Field Marshal the Viscount Slim KG GCB GCMG GCVO DSO MC*, London: Leo Cooper, 1976.

Liddell Hart, Captain B. H. *A Science of Infantry Tactics Simplified*, London: William Clowes, 1923.

Lind, W. S. *Manoeuvre Warfare Handbook*, London: Westview Press, 1985.

Lindsay, Martin. *So Few Got Through. With the Gordon Highlanders from Normandy to the Baltic*, Barnsley: Leo Cooper, 2000.

Luttwak, Edward and Horowitz, Dan. *The Israeli Army*, London: Allen Lane, 1975.

Luttwak, Edward. *Strategy. The Logic of War and Peace*, Cambridge, MA: Harvard University Press, 1987.

Manstein, Field Marshal Erich von. *Lost Victories*, London: Methuen, 1994.

Manstein, Rudiger von and Fuchs, Theodor, *Soldat im 20. Jahrhundert*, Munich: Bernard & Graebe, 1981.

March *et al. Decisions and Organizations*, Oxford: Basil Blackwell, 1988.

Marshall, S. L. A. *Armies on Wheels*, London: Faber & Faber, 1942.

Battles Hitler Lost. First-Hand Accounts of WW2 by Russian Generals on the Eastern Front. [Each battle described by a Marshal of the Soviet Union] New York: Richardson & Steinman, 1986.

McPherson, James M. *The Battle Cry of Freedom. The Civil War Era*, New York: McGill, 1987.

Mellenthin, Major-General F.W. von, trans. H. Betler. *Panzer Battles. A Study of the Employment of Armour in the Second World War*, London: Futura, 1979.

Middlebrook, Martin. *The Kaiser's Battle*, London: Penguin, 1983.

Morgan, Hugh and Seibel, Jürgen. *Combat Kill*, Cambridge: Patrick Stephens, 1997.

Morris, T. and Summers, J. (eds). *Sports Psychology: Theory, Applications and Issues* (2nd edn), Chichester: Wiley, 2002.

Moynahan, Brian. *The Claws of the Bear*, London: Hutchinson 1989.

Myers, Katherine D. and Kirby, Linda K. *Introduction to Type Dynamics and Development. Exploring the Next Level of Type*, Oxford: Oxford Psychologists Press, 2000.

Nafziger, George F. *The German Order of Battle. Panzers and Artillery in World War II*, London: Greenhill Books, 1998.

Nafziger, George F. *The German Order of Battle. Infantry in World War II*, London: Greenhill Books, 2000.

North, John. *Northwest Europe 1944–5. The Achievement of 21st Army Group*, London: HMSO, 1953.

Nosworthy, Brent. *The Battle Tactics of Napoleon, his Armies and Enemies*, London: Constable, 1993.

O'Ballance, Edgar. *No Victor, No Vanquished. The Yom Kippur War*, Novato, CA: Presidio Press, 1978.

O'Gorkiewicz, R. M. *The Technology of Tanks*, Coulsdon: Jane's, 1991.

Oman, Sir Charles. *A History of the Art of War in the Sixteenth Century*, London: Greenhill Books, 1987.

Oman, Sir Charles. *A History of the Art of War in the Middle Ages*, London: Greenhill Books, 1991.

The Concise Oxford Dictionary, 6th edn, Oxford: Oxford University Press, 1976.

Paret, P. (ed.). *Makers of Modern Strategy*, Oxford: Oxford University Press, 1986.

Patton Jr, George S. *War as I Knew it*, Cambridge, MA: Houghton Mifflin, 1947.

Perret, Bryan. *A History of Blitzkrieg*, New York: Stein & Day, 1983.

Piekalkiewicz, Janus, trans. H. A. and A. J. Barker. *Arnhem 1944*, London: Ian Allan, 1977.

Popkin, Richard H. and Stroll, Avrum. *Philosophy*, Oxford: Heinemann, 1989.

Popper, Karl. *The Logic of Scientific Discovery*, London: Hutchinson, 1959.

Province, Charles M. *Patton's Third Army*, New York: Hippocrene, 1992.

Pugh, D. S. (ed.). *Organization Theory*, Harmondsworth: Penguin, 1971.

Rauss, Generaloberst Erhard and Natzmer, Generalleutnant Oldwig von, trans. Peter G. Tsouras. *The Anvil of War. German Generalship in Defense on the Eastern Front*, London: Greenhill Books, 1994.

Reynolds, Michael. *Steel Inferno. 1st SS Panzer Corps in Normandy*, Staplehurst: Spellmount, 1997.

Roberts, Major-General G.P.B., CB DSO MC. *From the Desert to the Baltic*, London: William Kimber, 1987.

Roebling, Karl. *Great Myths of WW2*, New York: Paragon Press, 1985.

Rohwner, Jürgen. *Axis Submarine Successes 1939–45*, Cambridge: Patrick Stephens, 1983.

Roscoe, Theodore. *US Submarine Operations in WW2*, Annapolis, MD: US Naval Institute, 1949.

Rosinski, Herbert. *The German Army*, London: Pall Mall Press, 1966.

Rowland, David. *The Stress of Battle. Quantifying Human Performance in Combat*, Norwich, The Stationery Office, 2006.

Samuels, Martin. *Doctrine and Dogma*, New York: Greenwood, 1992.

Seaton, Albert. *The German Army 1933–45*, London: Weidenfeld & Nicolson, 1982.

Senich, Peter R. *The German Sniper 1939–45*, London: Arms & Armour Press, 1982.

Sheffield, Gary and Todman, Dan (eds). *Command and Control on the Western Front. The British Army's Experience 1914–18*, Staplehurst: Spellmount, 2004.

Sims, Edward H. *The Fighter-pilots*, London: Corgi, 1970.

Smith, John P. *Airborne to Battle. A History of Airborne Warfare 1918–71*, London: William Kimber, 1971.

Smith, General Rupert. *The Utility of Force. The Art of War in the Modern World*, London: Allen Lane, 2005.

Smithers, A. J. *Rude Mechanicals. An Account of Tank Maturity during the Second World War*, London: Grafton, 1987.

Spick, Mike. *The Ace Factor: Air Combat and the Role of Situational Awareness*, Shrewsbury: Air Life, 1988.

Spiller, Roger J. (ed.). *Combined Arms in Action since 1939*, Fort Leavenworth, KA: US Command and General Staff College Press, 1992.

Stanton, Shelby. *The Rise and Fall of an American Army. US Ground Forces in Vietnam*, Novato, CA: Presidio Press, 1985.

Strachan, Hew. *European Armies and the Conduct of War*, London: George Allen & Unwin, 1983.

Strauss, Franz Josef. *Geschichte der 2. (Wiener) Panzer Division*, Friedberg: Podzun-Pallas, 1987.

Stumpf, Reinhard. *Die Wehrmacht-Elite. Rang- und Herkunftsstruktur der deutschen Generale und Admirale, 1933–45*, Boppard am Rhein: Harald Boldt, 1982.

Sulzbach, Herbert, trans. Richard Thonger. *With the German Guns. Four Years on the Western Front*, Barnsley: Leo Cooper, 1998.

Swinson, Arthur. *The Raiders. Desert Strike Force*, Illustrated History of World War II, London: Pan/ Ballantine, 1968.

Tarrant, V. E. *The U-boat Offensive, 1914–45*, London: Arms & Armour Press, 1989.
Terraine, John. *The Smoke and the Fire. Myths and Anti-Myths of War 1861–1945*, London: Book Club Associates, 1981.
Tien, Chen-Ya. *Chinese Military Theory – Ancient and Modern*, Stevenage: Spa Books, 1992.
Townshend, Charles (ed.). *The Oxford Illustrated History of Modern War*, Oxford: Oxford University Press, 1997.
Trythall, Anthony John. *'Boney' Fuller – The Intellectual General*, London: Cassell, 1977.
Tsouras, Peter G. *Changing Orders. The Evolution of the World's Armies, 1945 to the Present*, London: Arms & Armour Press, 1994.
Vallance, Andrew G. B. *The Air Weapon. Doctrines of Air Power Strategy and Operational Art*, Basingstoke: Macmillan, 1996.
Verbruggen, J. F. *The Art of Warfare in Western Europe During the Middle Ages*, Woodbridge, Suffolk: Boydell Press, 1997.
Vigor, P. H. *Soviet Blitzkrieg Theory*, Basingstoke: Macmillan, 1983.
Waldrop, M. Mitchell. *Complexity. The Emerging Science at the Edge of Order and Chaos*. London: Penguin, 1994.
Warnery, Major-General Emanuel von. *Remarks on Cavalry*, London: Constable, 1997.
Westphal, General Siegfried. *The German Army in the West*, London: Cassell, 1951.
Whitaker, Brigadier-General Denis *et al. Victory at Falaise. The Soldier's Story*, London: HarperCollins, 2000.
White, Gilbert. *The Natural History of Selbourne*, Harmondsworth: Penguin, 1981.
Williams, Garnet P. *Chaos Theory Tamed*, London: Taylor & Francis, 1997.
Williams, John. *The Ides of May. The Defeat of France May–June 1940*, London: Constable, 1986.
Williamson, Gordon. *Aces of the Reich*, London: Arms & Armour Press, 1989.
Winton, Harold R. *To Change an Army. General Sir John Burnett-Stuart and British Armoured Doctrine 1927–38*, KA: 1988.
Wintringham, Tom and Blashford-Snell, J.N. *Weapons and Tactics*, London: Penguin, 1973.
Zetterling, Niklas and Frankson, Anders. *Kursk 1943, A Statistical Analysis*, London: Frank Cass, 2000.

2: BRITISH MILITARY PUBLICATIONS

Items are shown broadly in date order.

The Official History of the War. Military Operations in France and Belgium 1914–18. Volume IX, *The German March Offensive and its Preliminaries*. London: Macmillan, 1935.

Army Lists:
The Army List, April 1940.
The Army List, June 1943.
The Army List, January 1945.
The Army List, January 1946.
The Army List, Spring 1973.
The Army List 1995, Part II (biographical).

Establishment Tables:
II/100/3 (Headquarters of an Armoured Division) effective 30 November 1943.
II/104/3 (HQ Armoured Division Artillery) effective 8 December 1943.
III/181/2 (A Commander REME); effective 10 December 1943.
II/261/2 (Headquarters of a Commander, Royal Army Service Corps) effective 24 February 1943.

III/290/1 (Infantry and Armoured Divisional Ordnance Field Parks) effective 7 December 1943.

II/215/1 (Armoured Divisional Signals) effective 20 February 1945.

The Instructor's Handbook on Fieldcraft and Battle Drill (Provisional), Army Code 60314, issued under the direction of C.-in-C. Home Forces, apparently dated October 1942.

The Second World War 1939–45. Ordnance Services. The War Office, 1950.

The History of The Second World War. Army Medical Services – Administration. Volume One. London: HMSO, 1953.

The Second World War 1939–45. Army Signal Communications. The War Office, 1954.

The Second World War 1939–45. Army Supply and Transport. The War Office, 1954.

Order of Battle. The Second World War 1939–45. London: HMSO, 1960.

Command of Armour in World War II. Part 1: *North Africa*; Part 2: *North West Europe*. Films: British Defence Film Library C1404, C1405, 1979.

Design for Military Operations. The British Military Doctrine. Army Code 71451, 1989.

Army Doctrine Publication (ADP) Volume 1 Operations. Army Code 71565, 1994.

ADP Volume 2 Command. Army Code 71564, 1995.

Army Field Manual (AFM) Volume 1 Part 1, Formation Tactics. Army Code 71587, 1996.

AFM No 1 The Fundamentals. Army Code 71622, 1998.

Design for Military Operations. The British Military Doctrine. (2nd edn) Army Code 71451, 1996.

AFM Volume IV Pt 5, Operations in Built-Up Areas. Army Code 71657, 1998.

Staff Officer's Handbook. Army Code 71038, July 2001.

The Army Force Development Handbook, D/DGD+D/124/12/LW4, dated March 1996.

UK Doctrine for Joint and Multinational Operations. Joint Warfare Publication 0–10, 1st edn, September 1999.

Glossary of Joint and Multinational Terms and Definitions. Joint Warfare Publication 0–01.1, April 2001.

Officer Career Development. Unreferenced army publication dated 30 July 2001.

Operations in Iraq. An Analysis from a Land Perspective. Army Code 71816 (undated, but published in 2003).

3: US ARMY PUBLICATIONS

Items are shown in date order.

Sixth [US] Armored Division 1944–5. Combat Record of the Sixth Armored Division in the European Theatre of Operations, 18 July 1944 – 8 May 1945. Compiled under the Direction of Major Clyde J. Burke, Asst G3. Steinbeck-Druck, Aschaffenburg, undated. (Foreword dated 30 July 1945).

US Army Order of Battle WW2. European Theatre, Divisions. Office of the Theatre Historian, Paris, 15 December 1945.

The Organization of Ground Combat Troops. Office of the Chief of Military History, United States Army, Department of the Army, Washington, DC, 1947.

The Procurement and Training of Ground Combat Troops. Office of the Chief of Military History, United States Army, Department of the Army, Washington, DC, 1948.

Night Combat. Department of the Army, Pamphlet 20–236, June 1953.

The German General Staff Corps. Military Intelligence Division, War Department, Washington, DC, April 1946.

The Principles of Employment of Armor. Special Text No. 48, The Armor School, 1948, in *Armor*, May–June 1998.

Combat in Russian Forests and Swamps. Department of the Army Pamphlet 20–231, July 1951.

Field Manual (FM) 100–5, Fighting Future Wars. Department of the Army, Brassey's, McLean, VA, 1994.

FM 101–5, Staff Organizations and Operations. HQ, Department of the Army, Washington, DC, June 1999.

Maps Series M709, Editions 3–DMA, US National Imaging and Mapping Agency, Sheets 3079/I, 3080/ II–III, 3178/IV, 3179/III–IV.

FM 3–24 and Fleet Marine Force Manual 3–24, Counterinsurgency. HQ, Department of the Army, Washington, DC, December 2006.

4: THESES, PAPERS AND PRESENTATIONS

Items are shown in date order.

Slim, F.M. Sir William. Higher Command in War. Address to US Army Command and General Staff Course, 8 April 1952, *Military Review*, LXX.5, May 1990.

Manstein, F.M. Erich von. Grundsätzliche Gedanken an der Organizationsplan Heer, November 1955. *Soldat im 20. Jahrhundert*, Rudiger von Manstein and Theodor Fuchs, Munich 1981, trans. Ghislaine Fluck, with my assistance.

Command and Control in the Wehrmacht, unsigned (British Army) Staff College paper, 11 July 1985, Tactical Doctrine Retrieval Cell (TDRC), item no 7231.

Rowland, D. 'The Use of Historical Data in the Assessment of Combat Degradation', *Journal of the Operational Research Society*, 38.2, 1987.

Mackay-Dick, MBE, Brigadier I. C. *The Desert War 1940–43. Are any of the Lessons Relevant to NATO's Central Front Today?*, HCSC 1989. TDRC item no. 9104, dated 3 April 1989.

Williams, OBE, Brigadier J.G. *Streamlining Headquarters*, 11 February 1992. TDRC item no. 10184, dated 12 March 1992.

Sale, Richard. 'Towards a Profile of the Successful Army Officer'. *Defence Analysis*, 8.1, 1992.

Storr, Major J.P. *et al. Officers' Analytic and Communication Skills*, Group Research Paper, ACSC 28, November 1994.

Rouse, Major J.F. 'Introducing the Military Hybrid Continuum. A Decision Method for the Future Manoeuvre Army', PhD thesis, Cranfield University (RMCS Shrivenham), 1996.

Lieutenant-General Paul van Riper, Commanding General of the US Marine Corps Combat Development Center, prepared testimony to the House of Representatives National Security Committee, March 1997.

Brewster Smith, M. 'The Authoritarian Personality: A Re-review 46 Years Later'. *Political Psychology*, 18.1, 1997.

Jaya-Ratnam, Dr D.D.J.J. 'Close Combat Suppression: Need, Assessment and Use'. *Proceedings of the 34th Annual Gun and Ammunition Symposium*, 26–29 April 1999, Monterey, CA.

Anthony, Robert W. 'Relating Large and Small in C2 and Operations'. *Proceedings of the Command and Control Research and Technology Symposium (Proc CCRTS)*, US Naval War College, Newport, RI, June 1999, pp. 650–9.

McCann, Carol and Pigeau, Ross. 'Clarifying the Concepts of Control and of Command'. *Proc CCRTS*, US Naval War College, Newport, RI, June 1999, pp. 475–90.

Kiszely, Major-General John, presentation at the RUSI (Battle Management Systems Symposium), 18–19 November 99. *Journal of the Royal United Services Institute for Defence Studies (JRUSI)*, 144.4, December 1999.

Storr, Major J.P. 'Command and Control within the Land Component', *Journal of Battlefield Technology*, 3.1, March 2000.

In Praise of Attrition. Inaugural Lecture by Christopher Bellamy as Professor of Military Sciences and Doctrine, Cranfield University, 14 June 2001.

Barry, Colonel P. G. 'The Head or the Heart – an Assessment of the Value of Psychometric Testing in the Selection of Army Officers Following Technical Staff Training', MDA thesis, Cranfield University (RMCS Shrivenham) 2001.

Storr, Jim. 'The Commander as Expert', Ch. 6 of *The Big Issue*, SCSI Occasional No. 45, 2002a.

Storr, Lieutenat-Colonel J.P. 'The Nature of Military Thought', PhD thesis, Cranfield University (RMCS Shrivenham), 2002b.

Aidman, Eugene and Schofield, Grant. 'Personality and Individual Differences in Sport'. T. Morris, and J. Summers, (eds), *Sports Psychology: Theory, Applications and Issues* (2nd edn), Milton: Wiley, 2002.

5: ARTICLES AND LETTERS

Items are shown in date order.

Lindsay, Lieutenant-Colonel (retd) Sir Martin of Dowhill Bt, CBE DSO. 'Thoughts on Command in Battle', *British Army Review (BAR)* 69, December 1981.

Taylor, CBE DSO, Brigadier (retd) George. 'Further Thoughts on Command in Battle', *BAR* 71, August 1982.

Lynam, Major J.M. 'Exercise King's Ride V: Initial Impressions', *Army Training News*, April 1986.

Spartacus. 'Simplicity', *BAR* 98, August 1991.

Rowland, D. 'The Effect of Combat Degradation on the Urban Battlefield', *Journal of the Operational Research Society, 42.7, 1991.*

Storr, Major J. P. 'FIBUA: The Tactics of Mistake', *Army Doctrine and Training News*, 1, 1994.

Corum, James S. 'From Biplanes to Blitzkrieg: The Development of German Air Doctrine Between the Wars', *War in History*, 3.1, 1996.

Spartacus. 'The Pretty Adjutant Syndrome', *BAR* 115, April 1997.

Gat, Azar. 'British Influence and the Evaluation of the Panzer Arm: Myth or Reality?' Part 1, *War in History*, 4.2, 1997: pp. 150–73.

Erskine, Freda 'Why People Stay Healthy', British Medical Journal 314, 3 May, 1997.

Irving, Mark. 'Expanding Universe', *Connected*, Supplement to the *Daily Telegraph*, 5 August 1997.

Antal, Lieutenant-Colonel John F. 'It's Not The Speed of the Computer that Counts! The Case For Rapid Battlefield Decision-making', *Armor*, May–June 1998.

Kiszely, Major-General John. 'The Meaning of Manoeuvre', *JRUSI*, December 1998.

Army Historical Branch letter HB(A) 6/3 dated 1 December 1998.

Caddick Adams, Captain Peter. 'The TA Before the Second World War', *BAR* 121, April 1999.

'The Warrior Class', *US News (and World Report)*, 5 July 1999.

Carver, Field Marshal Lord. 'The Boer War', *JRUSI*, 144.6, December 1999.

HB(A) Letter 11/8/1 dated 17 November 2000.

Welch, Michael. 'The Centenary of the British Publication of Jean de Bloch's "Is War Now Impossible?" (1899–1902)', *War in History*, 7.3, 2000.

Bennett, Matthew. 'The Crusaders' "Fighting March" Revisited', *War in History*, 8.1, 2001.

Stuttaford, Dr Thomas. 'Telltale Signs of a Textbook Personality Disorder', *The Times*, 29 January, 2001.

Storr, Major Jim. 'A Year Observing Command and Control', *BAR* 126, Spring 2001.

Nicol, Captain J. D. 'The Morale of the Australian Infantry in South Vietnam, 1965–72', *BAR* 127, summer 2001.

Bird, Tim. 'UK Effects-Based Planning and Centre of Gravity Analysis: An Increasingly Dysfunctional Relationship?', *JRUSI*, 153.2, April 2008.

6: REPORTS

Items are shown in date order.

Effect of Artillery Fire in Attacks in Mountainous Country, (British) No. 1 Operational Research Section report 1/24/A, 1945.

The Contribution of Manned Armoured Reconnaissance to the Information Gathering Assets of 1 (BR) Corps – Exercise White Ermine, Final Report, DOAE Study 426, MOD Defence Operational Analysis Establishment Memorandum R9003, dated June 1991.

The Effects of Shock and Surprise on the Land Battle, MOD Defence Operational Analysis Establishment Memorandum R9301.

Rowland, D. *et al. Breakthrough and Manoeuvre Operations – Historical Analysis of the Conditions for Success*, MOD Defence Operational Analysis Centre Report R9412, October 1994.

Marsden, J. L. *Span of Command and Control in the British Army: An Initial Study*, DERA Farnborough, DRA/CHS/HS3/CR96026/2.0, dated August 1996.

Land Command Observations from Training, HQ Land Command, 1997.

Rooney, D. *The Potential for Representing Combat Participation in Operational Analysis Models*, DERA/CHS/HS3/CR97O11/1.0, March 1997.

The 20th Century Variation of C2 Over Time, CDA/HLS/PR179/E5, dated 19 December 1997.

Exercise Sea Wall. Headquarters Infantry file 109/00/01, dated 30 December 1997.

Rooney, D. *et al. Mission Command and Battlefield Digitization: Human Sciences Considerations*, DERA CHS/HS3/CR980097/1.0, dated March 1998.

Storr, Major J. P. *The Attack on Will at Low Level*, DERA CDA, DERA/CDA/HLS/WP00067, dated April 2000.

Report of the Committee on the Lessons of the Great War (The Kirke Report). *BAR* Special Edition, April 2000.

ISTAR Integration Warfighting Experiments 2002. QINETIQ/CONSULT/CR036134/1.0, dated June 2003.

Index

Notes:

1. Page references in **bold** refer to Figures and Tables.
2. Military personnel are given the highest rank cited in the text.

Abraham, Maj Gen M. 166
abstraction 53, 136
ace pilots 13, 14, 159–60, 162
'Achilles and the Tortoise' 17
Adam, Lt Gen R.F. **158**
Afghanistan 1, 36
aggressive exploitation 98
aggressive ground
   reconnaissance 107, 109,
   114, 125
      effectiveness 50–1, 56, 115
agility 117, 124
aims and means (Clausewitz's
   Dialectic) 59–61, **60**, 70, **71**,
   92, 98, 99, 103
air superiority 49–50
airpower 68, 69–70, 102, 115
Alanbrooke, Field Marshal
   The Viscount 168
Alexander, Maj Gen H.R. 155,
   158, 168
Allenberger, S. 157, 161
Amer, Field Marshal A.H. 82
Andropov, Y. 168
Another Bloody Century
   (Gray) 2
antiaircraft units 55–6, 69,
   74, 115
antitank weapons and units
   effectiveness 68, 69, 77
   use 55, 74, 90, 115, 125–6
Arab armies 55–6, 167
armies
   organization of 115–17
   see also organization of
      forces
   peacetime development
   187–8, 195–6
armoured brigades 126–7

armoured divisions
   optimum size 125
   Second World War and
      modern compared
      115–17, 118
armoured reconnaissance
   units 127
armoured warfare 55–6
   see also tanks
articulation 74–5, 88, 117
artillery
   logistic requirements 125
   neutralization 83, 87, 88
   suppression 89, 90, 108
   troop type model 68–9, **69**
   use 97, 100, 108, 115, 125
asymmetry 104–5
Atkins, Tommy 205
'attack is the best form of
   defence' 16, 17, 93–4, 103,
   104, 198
attack tactics 107–10
attrition 53–4, 61, 73, 94
Auerstadt, Battle of 187
Australian Army 191
Austro-Prussian War 32
authoritarians
   characteristics 176–8, 185,
   186
   in combat situations 177,
   178, 183
   in peacetime 176, 178,
   184, 186
   senior officers as 178–81,
   179, 183, 195
autocrats
   characteristics 176, 177
   in combat situations 183,
   184

   senior officers as 178–81,
   183
autonomous action 75

Bacon, F. 25, 27
Bagnall, Field Marshal Sir N.
   61, 167, 189
Balck, Col W. 95
Bannockburn, Battle of 65
BAOR *see* British Army of the
   Rhine
Barker, Lt Gen M.G.H. **158**
Barry, Col P. 181–2, 183, 185
Bartholomew Committee
   116
battalions
   modelling 45, **46**
   utilization 120, **121**
Battle of the Atlantic 36
Battle of Britain 163
Battle of the Bulge 80, 126
Beattie, Adm Sir D. 176
Beck, Gen L. 192
Berkeley, G. 25
Bertalanffy, L. von 46, 61
Blumenson, M. 178
Boer War 156, 168, 205
Bonaparte, Napoleon *see*
   Napoleon I
Bosnia 36, 185
Boyd, Col J. 13
Bradford, Capt B. 213n 69
Brezhnev, L. 168
brigades
   organization and size 15,
   118, 122, 124
   planning time 134,
   145–6
   support 125, 126–7

Lightning Source UK Ltd.
Milton Keynes UK
UKOW030830130112

185309UK00001BA/14/P